FREEZE FRACTURE:
METHODS, ARTIFACTS, AND INTERPRETATIONS

Freeze Fracture:
Methods, Artifacts, and Interpretations

Editors

John E. Rash, Ph.D.
Department of Anatomy
Colorado State University
Fort Collins, Colorado

C. Sue Hudson, Ph.D.
Department of Pharmacology
  and Experimental Therapeutics
University of Maryland
School of Medicine
Baltimore, Maryland

Raven Press ■ New York

Raven Press, 1140 Avenue of the Americas, New York, New York 10036

Made in the United States of America

Library of Congress Cataloging in Publication Data
Main entry under title:

Freeze fracture.

Includes bibliographical references and index.
1. Freeze fracturing. 2. Membrane (Biology)
I. Rash, John E. II. Hudson, C. Sue. [DNLM:
1. Freeze fracturing. QH236.2 F857]
QH236.2.F74   574.8'75'028   79–63036
ISBN 0–89004–386–8

Second Printing, May 1981

Preface

During the past 5 years, the freeze-fracture/freeze-etch technique has become the primary morphological technique for analyzing membrane structure, and promises in the very near future to provide methods for visualizing, identifying, and quantifying *in situ* individual classes of transmembrane and nonmembrane-bound macromolecules. The wealth of new biological information obtained by freeze-fracture techniques has resulted in a quantal jump in our understanding of cell structure and function at the molecular level and demands that every cell biologist become familiar with the technique and its interpretation.

Freeze Fracture: Methods, Artifacts, and Interpretations represents the concerted efforts of representatives from most of the major laboratories using freeze-fracture and freeze-etch methodologies. It provides a logical, step-by-step introduction to the freeze-fracture technique, as well as a useful guide for recognizing and minimizing the major artifacts and preparative faults commonly encountered. New techniques and improved procedures are described and old problems are presented in a new setting. Unresolved problems and current controversies are presented and avenues for future experimentation are suggested. We are particularly pleased because this volume clearly reflects the continuing cooperation of the many freeze-fracture laboratories and provides evidence for a genuine and abiding desire by each of the laboratories represented to improve the quality and interpretation of freeze-fracture replicas.

We have enjoyed organizing the text and have benefited significantly from many of the observations and improved procedures. We hope that this volume provides the necessary impetus to encourage biologists in diverse disciplines to explore that fascinating region where structure and function merge at the molecular level.

John E. Rash
C. Sue Hudson

Acknowledgments

We gratefully acknowledge the American Society for Cell Biology for hosting the original symposium, the numerous participants in the symposium, and the many additional contributors to this book.

For conscientious and rapid review of manuscripts, we express our sincere appreciation to Drs. Russell L. Steer, L. Andrew Staehelin, and Ronald Weinstein. We also thank Mr. Rick Wierwille, Mr. William F. Graham, and Mrs. Laura L. Rash for assistance in preparing the text.

Finally, we are indebted to the editorial staff at Raven Press for the continuing support that ensured the very rapid publication of these timely reports.

Contents

Contributors

David J. Benefiel
Department of Anesthesia
San Francisco General Hospital
University of California
San Francisco, California 94110

S. Böhler
Balzers, Furstentum
Liechtenstein

D. Branton
Cell and Developmental Biology
The Biological Laboratories
Harvard University
Cambridge, Massachusetts 02138

R. Malcolm Brown, Jr.
Botany Department 010-A
University of North Carolina
Chapel Hill, North Carolina 27514

Stanley Bullivant
Department of Cell Biology
University of Auckland
Auckland, New Zealand

Douglas E. Chandler
Department of Physiology
University of California
San Francisco, California 94143

A. Elgsaeter
Institute for Biophysics
University of Trondheim
N-7034 Trondheim-NTH, Norway

Mark H. Ellisman
Department of Neurosciences
University of California, San Diego
La Jolla, California 92093

Eric F. Erbe
Plant Virology Laboratory
Plant Protection Institute, AR, SEA
U.S. Department of Agriculture
Beltsville, Maryland 20705

William F. Graham
Department of Pharmacology and
    Experimental Therapeutics
University of Maryland
School of Medicine
Baltimore, Maryland 21201

Heinz Gross
Institut für Zellbiologie
Eidgenössische Technische Hochschule
    (ETH)
Hönggerberg, 8093 Zürich, Switzerland

David L. Hasty
Department of Anatomy
Harvard Medical School
Boston, Massachusetts 02115

Elizabeth D. Hay
Department of Anatomy
Harvard Medical School
Boston, Massachusetts 02115

C. Sue Hudson
Department of Pharmacology and
    Experimental Therapeutics
University of Maryland
School of Medicine
Baltimore, Maryland 21201

Stefan Kirchanski
Cell and Developmental Biology
The Biological Laboratories
Harvard University
Cambridge, Massachusetts 02138

Gerd G. Maul
The Wistar Institute of Anatomy and Biology
Philadelphia, Pennsylvania 19104

Peter Metcalf
Department of Cell Biology
University of Auckland
Auckland, New Zealand

Kenneth R. Miller
The Biological Laboratories
Harvard University
Cambridge, Massachusetts 02138

J. Michael Moseley
Plant Virology Laboratory
Plant Protection Institute, AR, SEA
U.S. Department of Agriculture
Beltsville, Maryland 20705

W. E. Mowczko
Department of Health, Education, and Welfare
Public Health Service
National Institutes of Health
Bethesda, Maryland 20014

Richard Ornberg
Section Functional Neuroanatomy
National Institute of Neurological and Communicative Disorders and Stroke
National Institutes of Health
Building 36, Room 3B-24
Bethesda, Maryland 20014

Bendicht U. Pauli
Department of Pathology
Rush Medical College and Rush-Presbyterian–St. Luke's Medical Center
Chicago, Illinois 60612

Karl H. Pfenninger
Department of Anatomy
Columbia University
New York, New York 10032

Pedro Pinto da Silva
Section of Membrane Biology
Laboratory of Pathophysiology
National Cancer Institute
National Institutes of Health
Bethesda, Maryland 20205

John E. Rash
Department of Pharmacology and Experimental Therapeutics
University of Maryland
School of Medicine
Baltimore, Maryland 21201

Thomas S. Reese
Section on Functional Neuroanatomy
National Institute of Neurological and Communicative Disorders and Stroke
National Institutes of Health
Bethesda, Maryland 20205

Birgit Satir
Department of Anatomy
Albert Einstein College of Medicine
Yeshiva University
Bronx, New York 10461

Peter Satir
Department of Anatomy
Albert Einstein College of Medicine
Yeshiva University
Bronx, New York 10461

Joel B. Sheffield
Department of Biology
Temple University
Philadelphia, Pennsylvania 19122

Emma Shelton
Department of Health, Education, and Welfare
Public Health Service
National Institutes of Health
Bethesda, Maryland 20205

L. Andrew Staehelin
Department of Molecular, Cellular, and
*   Developmental Biology*
University of Colorado
Boulder, Colorado 8030y

Russell L. Steere
Plant Virology Laboratory
Plant Protection Institute, AR, SEA
U.S. Department of Agriculture
Beltsville, Maryland 20705

Kennedy P. Warne
Department of Cell Biology
University of Auckland
Auckland, New Zealand

Ronald S. Weinstein
Department of Pathology
Rush Medical College and Rush-
*   Presbyterian-St. Luke's Medical Center*
Chicago, Illinois 60612

J. H. Martin Willison
Biology Department
Dalhousie University
Halifax, Nova Scotia, B H 4J1 Canada

Freeze Fracture: Methods, Artifacts, and Interpretations, edited by J. E. Rash and C. S. Hudson. Raven Press, New York © 1979.

Introduction to Sample Preparation for Freeze Fracture

C. Sue Hudson, John E. Rash, and William F. Graham

Department of Pharmacology and Experimental Therapeutics, University of Maryland School of Medicine, Baltimore, Maryland 21201

The relatively recent development of freeze-fracture methodologies as practical research tools for splitting and high resolution replication of lipid bilayers has provided several new and unique means for the analysis of the macromolecular architecture of biological membranes. In this report, which is designed to assist the novice, we describe our standard regimen for replica preparation. However, freeze fracture is a sophisticated technique requiring careful attention to a variety of details and disciplines including the physical and chemical properties of water, lipids, and proteins, their interactions at ultralow temperatures, and the properties of each under conditions of molecular shear. Furthermore, the investigator must have at least a limited understanding of the properties of vaporized metals under high vacuum, their codeposition with vaporized carbon on ultracold surfaces of varying composition, the possible interaction or alteration of the deposited metals with the surfaces during melting and after tissue thawing, and possible changes of certain shadowing metals during cleaning with harsh reagents (1,2, 4,5,7,8). Therefore, it is advised that anyone wishing to enter the field do so under the continuing close supervision of an experienced investigator.

The steps to be considered when preparing conventional replicas are ultrarapid freezing, chemical fixation, cryoprotection, freezing, cleaving, etching, replication, cleaning, and mounting. Some steps are mutually exclusive, whereas others may be eliminated or modified because of the unique qualities of some tissues. Therefore, it should be emphasized at the outset that the investigator keep a detailed record of each preparative step for each sample. This record is invaluable in determining the source of artifact if a replica is of unsatisfactory quality. Since each step of the preparative procedure possesses potential pitfalls for the novice, each is considered separately.

ULTRARAPID FREEZING

Cellular processes involving physical movement of membranes (e.g., membrane fusion during exocytosis, transmitter release, or fertilization) may not be arrested sufficiently rapidly by formaldehyde or glutaraldehyde, fixatives that do not fix

1

or immobilize lipids. However, since lipid fixatives such as osmium tetroxide and potassium permanganate alter the hydrophobic properties of the lipids, they cannot be employed for freeze-fracture. Furthermore, existing evidence indicates that many plant and animal cell membranes undergo major changes in particle partition coefficients after glytaraldehyde fixation (B. Satir and P. Satir; J. H. M. Willison and R. M. Brown, *this volume*), and that other membranes develop characteristic artifactual alterations (ref. 3; E. D. Hay and D. L. Hasty, *this volume*). The alternative approach in such cases is ultrarapid freezing of living material to halt activity almost instantaneously.

Ultrarapid freezing may be accomplished by one of two methods. Spray freezing (9) is limited to cellular or subcellular suspensions, monolayers, or surface cells of larger samples, whereas contact freezing against a supercooled copper block (4) is restricted to monolayers or thin regions on the surface of tissues. General use of the latter technique is further limited by the complexity of the equipment and the constraints of working with liquid helium. This technique is described in greater detail in the chapters by D. E. Chandler *(this volume)* and R. Ornberg and T. Reese *(this volume)*. For the novice, we note that unless a specific specimen demands the use of ultrarapid freezing, aldehyde fixation is generally employed.

CHEMICAL FIXATION

In the preparation of freeze-fracture samples, the primary goal of fixation is to immobilize subcellular and macromolecular structures very nearly as they exist in the living condition. Therefore, it is essential that the fixative reach and penetrate each cell as rapidly as possible. Since penetration of the primary fixative is limited primarily by diffusion, fixation by immersion is considered adequate only for the preparation of single cells, cell monolayers, very small specimens, or surface cells of larger specimens that cannot be perfused (e.g., biopsy material). Large blocks of tissue, particularly the highly vascularized organs and tissues of warm blooded animals with high metabolic rates, must be prepared by vascular perfusion fixation. Because diffusion of fixatives from the perfused capillaries to most cells occurs over micron distances, the quality of preservation of the tissue is improved significantly, presumably because the periods of cellular hypoxia and exposure to altered ion concentrations are greatly diminished or eliminated.

We routinely prepare a variety of rat, mouse, and chicken tissues by ventricular whole-body perfusion. To clear the blood and prevent clotting, oxygenated Ringer's solution containing 10 units/ml heparin is used as perfusate, followed immediately by 2.5% glutaraldehyde in the appropriate Ringer's solution. With this simple and rapid technique, anoxic changes in mitochondria (10) and rearrangement of membrane macromolecules appear to be minimized. The initial temperature of the perfusate must be at or near the *in vivo* temperature to preclude internal rearrangement and phase separation of membrane macromolecules (Fig. 1a; see also G. G. Maul, *this volume*). However, once the tissue is adequately

FIG. 1.a: Neuromuscular junction from a rat muscle maintained for 2 hr at 21°C, then fixed in 2.5% glutaraldehyde. The highly fluid nerve terminal membrane of mammals is very susceptible to lipid–protein phase separation artifacts. Note the particle-free domains and strings of particles *(arrow)* (×17,500). **b:** Plasma membrane of rat muscle fixed briefly with 2.5% glutaraldehyde and then stretched to 140% of rest length. "Stretch artifact" is resolved as particle-free "rivulets" separating particle-rich "islands" ×67,500.

preserved (usually within a few minutes), the specimen temperature should be reduced to and maintained at 0 to 4°C, primarily to prevent leaching of membrane lipids. To emphasize this point, we note that satisfactory replicas of muscle fiber membranes have been prepared from tissues stored in 2.5% buffered glutaraldehyde at 0°C for 3 to 4 months.

During the initial preparation of tissue by chemical fixation, additional factors may affect membrane integrity and macromolecular architecture. When dissecting fixed tissue, precautions should be taken to minimize physical trauma because partially fixed membranes are particularly vulnerable to "stretch artifacts" (Fig. 1b). Finally, the effects of membrane-active drugs such as local anesthetics and barbiturates (11), altered concentrations of sodium, potassium, calcium, magnesium, chloride, and phosphate ions, and exposure to such lipid solvents as ether and glycerol have not been well characterized. Thus, care should be taken to maintain all ions at or near extracellular levels in the *in vivo* preparations.

CRYOPROTECTION

Unless ultrarapid freezing techniques are employed, the freezing process creates damage in the specimens because of the formation of large ice crystals. Thus, cryoprotectants such as glycerol or dimethylsulfoxide (DMSO) are used to prevent the formation of detectable ice crystals. In some cases, however, natural cryoprotection may be provided by the extremely low water content of some plant tissues such as seeds.

For cryoprotection, tissues are microdissected (0.5 mm × 1 mm × 1 mm), and the specimens are placed in a small vial containing buffered fixative. A 30% glycerol solution (prepared with buffer or distilled water, depending on specimen requirements) is added dropwise with continuous mixing until the solution surrounding the tissue contains 5 to 10% glycerol. (We presume but have not adequately assessed the necessity for gradual initial infiltration.) Subsequently, multidrop additions of 30% glycerol are acceptable, but care should be taken to ensure adequate mixing after each addition. Infiltration to 30% glycerol usually requires about 15 min and is followed by equilibration in 30% glycerol for 30 to 45 min.

FREEZING

Freezing is conventionally performed in liquid Freon 12 (dichlorodifluoromethane), Freon 22 (monochlorodifuoromethane), or in a liquid nitrogen slurry. (Samples cannot be frozen rapidly in liquid nitrogen, since it is at its boiling point at −196°C and the heat in a specimen is sufficient to boil the nitrogen and form an insulating layer of gaseous nitrogen.) For convenience and because of slightly greater heat capacity of Freon 12 as compared to Freon 22, we routinely use commercially available sources such as Effa Duster. In the freezing process employed in our laboratory (Fig. 2a), two liquid nitrogen-containing

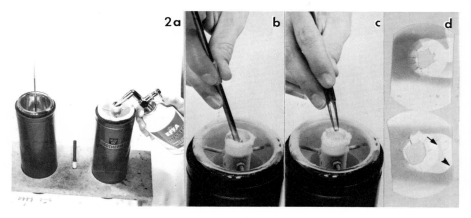

FIG. 2. a–c: Simple technique for freezing specimens. **d:** Filter paper shadow records of platinum *(arrow)* and carbon *(arrowhead)* layers. Note the very thin carbon layer (50 to 75 Å) in the lower shadow record.

Dewar flasks are utilized. The first contains the freezing unit, a small metal well containing liquid Freon 12 maintained at its freezing point by the surrounding bath of liquid nitrogen. The second contains a specimen storage basket immersed in liquid nitrogen. The freezing well is filled initially by gentle release of gaseous Freon 12 into the bottom of the precooled well (Fig. 2a). To avoid spraying liquified Freon out of the completely filled well, the nozzle is quickly removed from the well *while* Freon gas is being released. While the well of Freon freezes, glycerinated samples are poured into a petri dish and mounted on gold specimen supports using a dissecting microscope. If standard Balzers specimen supports with depression are used, they are submerged in 30% glycerol and cleared of all air bubbles which might be trapped in the support depression. (Changes in volume of bubbles frozen in the depression may dislodge the specimen during freezing or when the specimen warms to near − 100°C on the specimen stage.) If samples are extremely small, the depression may be filled with a few fibers of tissue paper soaked in 30% glycerol. After the sample is positioned on the specimen support, excess glycerol is blotted from the support and specimen. A metal thawing bar is used to thaw the entire central portion of the Freon well (Fig. 2b). When refreezing of the well is near completion, the remaining liquid Freon is near − 150°C. Then the mounted specimen is plunged into the Freon and held for 2 or 3 sec near the interface of the solid and liquid phases (Fig. 2c). The frozen sample is transferred quickly to the specimen basket in the second Dewar flask. Prechilled forceps should *not* be used to pick up samples *before* they are frozen. This can lead to freezing phase-separation artifacts because of gradual thermal changes occurring immediately prior to freezing. However, once a sample has been frozen, forceps tips must be chilled in liquid nitrogen before picking up the specimen. Once frozen, samples can be prepared for immediate cleaving or they can be transferred to prechilled specimen storage containers and stored in liquid nitrogen. Although the time limitation for storage has not been established, we have found excellent preservation of samples stored up to 3 years. However, these samples appear to be more brittle and have a greater tendency to crack during cleaving. Finally, should a sample become thawed or warmed above − 80°C, it need not be discarded, for we have observed that many such samples may be recovered by reinfiltration with 30% glycerol for 30 min before refreezing.

CLEAVING AND REPLICATION

Several precautions are necessary when preparing to cleave and replicate samples. To prevent temperature fluctuation of the samples, the transfer from the storage container to the specimen basket and then to the stage must be performed as quickly as possible using only prechilled blunt-tipped forceps. The specimen stage is prechilled to − 150°C in a *clean, evacuated* bell jar. To minimize the formation of frost, the bell jar should be vented only with dry helium or nitrogen gas. Immediately upon venting the bell jar, the specimen stage is lightly sprayed with Freon gas. The liquid Freon slurry that forms ensures maximum heat transfer between the specimen and specimen stage.

The actual fracturing and replication of tissue should be done according to the manufacturer's instructions and modified according to personal experience. However, we have found that it is useful to cut a filter paper to fit at the base of the specimen clip (Fig. 2d) to aid in determining platinum and carbon thickness. In the Balzers 360 M Freeze-Etch device used in our laboratory, inexpensive brittle injector razor blades cleaned with acetone are used to initiate the cleavage plane. Although no experimental data are available, many investigators prefer moderately slow, shallow passes of the knife through the frozen tissue. To minimize surface contamination by water vapor (arising primarily from specimen chips on the knife edge), it is extremely important to leave the knife either directly above or at the maximum distance from the specimens after each knife pass (see J. E. Rash et al., *this volume*). After the final knife pass, the specimen may be allowed to etch (freeze etch) or may be immediately replicated (freeze fracture). Etching is accomplished by positioning the knife chuck directly above the specimen to serve as a cold trap for subliming water vapor. Since there are no etchable cryoprotectants currently known, the etching depth is severely limited in previously cryoprotected specimens. During conventional freezing, the solutions often separate into eutectic and hypoeutectic phases (Fig. 3). The salts, mobile proteins, and other molecules may become highly concentrated in the "eutectic" phase which is more resistant to etching than the ice crystals of the dilute "hypoeutectic" phase.

To prepare a freeze-fracture replica, platinum deposition by emission from a carbon arc or from an electron-beam gun should be initiated immediately

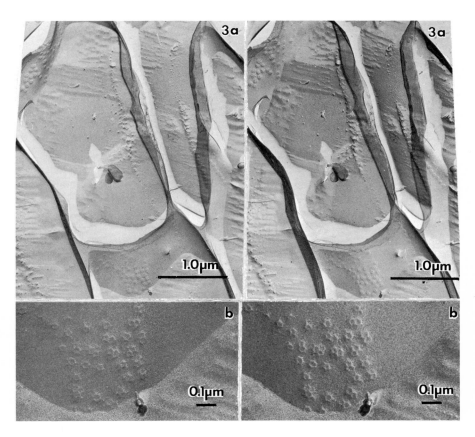

FIG. 3. a: A stereoscopic view of deep-etched purified coated vesicles from pig brain (from M. Woodward and J. Rash, *unpublished observations*). The coated vesicles are sequestered in the concentrated salt phase ("eutectic") which forms a nonetchable "honeycomb". ×18,000. **b:** High magnification view of the coated vesicles. ×56,000.

after the final cleave to minimize contamination (see J. E. Rash et al., *this volume*). Prior to carbon deposition, the knife is positioned immediately to the right of the cold stage so that it casts a shadow on the filter paper below (see Fig. 2d). Carbon deposition should follow immediately, primarily because deposition of water on the platinum layer prior to deposition of the carbon layer prevents proper stabilization of the replica. Later, during cleaning, partial dissolution of the platinum layer (Fig. 4) or separation of the carbon from the platinum layer may occur. It is useful to save the shadowed filter paper (Fig. 2d) and to record all relevant information such as vacuum conditions, platinum deposition time, cleaving patterns, and any other pertinent data. These data are often extremely useful in assessing the quality of the replica and determining the source of artifacts.

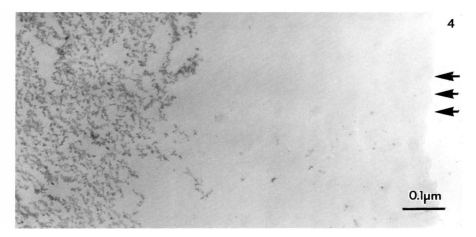

FIG. 4. A partially dissolved replica. A brief delay between platinum and carbon depositions presumably resulted in a discontinuous layer of water vapor which prevented proper stabilization of the platinum layer. Note the edge of the carbon support at arrows. ×112,000.

CLEANING AND MOUNTING REPLICAS

After replication, the stage temperature is lowered to − 150°C prior to venting the bell jar. After venting, any samples that were not cleaved are quickly returned to the storage basket. Cleaved specimens are picked up with room temperature forceps and placed on a finger tip to thaw. At this point, some investigators apply 0.5% collodion, which reduces the chance of breaking the replica. The collodion is removed after the cleaning process (C. F. Armstrong, *personal communication*). The specimen support plus tissue (Fig. 5a) is gently lowered at a slight angle into a bath containing the same solution used to infiltrate tissues prior to freezing (in this case, buffered 30% glycerol). The tissue plus replica usually separates from the support and floats on the surface of the solution. For this and subsequent cleaning steps, a white porcelain Coors Spot plate is used for transferring the replica through the cleaning solutions. Using a fine platinum loop, the specimen is transferred through 2-min changes in a series of decreasing glycerol concentrations (30%, 20%, 10%, 5%) and floated onto distilled water (Fig. 5b). To equilibrate the strong differences in surface tension, multiple-loop transfers of solution are always made before actually transferring the specimen to the next well. After three rinses in distilled water, the specimen is transferred to 1% sodium hypochlorite (20% bleach) for 30 min. This solution must be sufficiently dilute so that the tiny bubbles that form during digestion do not damage the replica. The replicas are cleaned in three successive changes of undiluted bleach (20 min each) and passed through several decreasing bleach concentrations to distilled water. After several distilled water rinses, the replica is transferred through increasing concentrations of chromic acid (10%, 20%, 30%, about 10 min in each bath) and transferred through three 20-min digestions

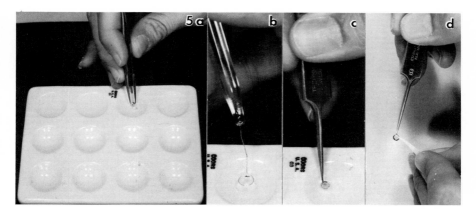

FIG. 5. a–d: Cleaning, transferring, and mounting of replicas. The replica is extremely fragile and easily damaged by direct physical contact. Air drying of the mounted replica can be accelerated by gently blotting the edge of the grid with a filter paper arrow.

in concentrated chromic acid (40%). (Some investigators use other agents such as sulfuric acid, sodium hydroxide, or solvents.) Rapid transfer through six to eight distilled-water baths (1 min each) and four to six additional distilled-water rinses (5 min each) removes residual chromic acid. It should be remembered that chlorine bleach is highly reactive with all acids, forming highly toxic chlorine gas. Therefore, even during disposal, the bleach and chromic acid solutions should not be mixed.

The cleaned replica is picked up on a standard 200- or 400-mesh electron microscope grid. [We have not yet assessed the advisability of formvar, collodion, or carbon support films, however, these are used routinely in some laboratories (R. L. Steere, *personal communication*).] To ensure that the replica adheres to the grid, precleaned grids are dipped into "grid glue" and placed on a filter paper to dry. (Grid glue is made by dipping approximately 2 in.^{2} of Scotch tape into 20 ml of ethylene dichloride for 10 sec.) To reduce repulsion of the replica by the hydrophobic grid surfaces, the grid is dipped into a dilute (1:30) solution of wetting agent (Photo-Flo 200) and two quick distilled-water rinses before being placed into the well containing the replica. After submersion of the grid, the replica is carefully centered and picked up from below (Fig. 5c). Excess water is gently blotted from between the tips of the forceps and from the *sides* of the grid only (Fig. 5d). If blotted directly on filter paper, the replica is destroyed by the rush of water through the grid squares. Replicas may be stored indefinitely.

SUMMARY

Freeze fracture is a rapidly evolving investigative technique requiring extensive background knowledge in diverse disciplines, meticulous attention to numerous details, and cautious interpretation of images. Thus, the successful freeze-fracture investigator continually strives to modify and improve current methods and attempts to devise new procedures that have the promise of improving the retrieval of useful biological information and minimizing artifactual or misinformation. It is hoped that this chapter provides the novice with a rational approach to this exciting new discipline.

REFERENCES

1. Bullivant, S. (1973): Freeze-etching and freeze-fracturing. In: *Advanced Techniques in Biological Electron Microscopy,* edited by J. K. Koehler, pp. 67–112. Springer-Verlag, New York.
2. Fisher, K., and Branton, D. (1974): Application of the freeze-fracture technique to natural membranes. In: *Methods in Enzymology, Vol. 32: Biomembranes Part B,* edited by S. Fleischer and L. Packer, pp. 35–44. Academic Press, New York.
3. Hasty, D. L., and Hay, E. D. (1978): Freeze-fracture studies of the developing cell surface. II. Particle-free membrane blisters on glutaraldehyde-fixed corneal fibroblasts are artefacts. *J. Cell Biol.,* 78:756–768.
4. Heuser, J. E., Reese, T. S., and Landis, D. M. D. (1976): Preservation of synaptic structure by rapid freezing. *Cold Spring Harbor Symp. Quant. Biol.* 40:17–24.
5. Koehler, J. K. (1968): The technique and application of freeze-etching in ultrastructure research. *Adv. Biol. Med. Phys.* 12:1–84.
6. Koehler, J. K. (1972): The freeze-etching technique. In: *Principles and Techniques of Electron Microscopy: Biological Applications,* Vol. 2, edited by M. A. Hayat, pp. 53–98. Van Nostrand Reinhold, New York.
7. Moor, H. (1966): Use of freeze-etching in the study of biological ultrastructure. In: *International Review of Experimental Pathology,* Vol. 5, edited by G. W. Richter and M. A. Epstein, pp. 179–216. Academic Press, New York.
8. Moor, H. (1969): Freeze-etching. In: *International Review of Cytology,* Vol. 25, edited by G. H. Bourne, J. F. Danielli, and K. W. Jeon, pp. 391–412. Academic Press, New York.
9. Plattner, H., Fischer, W. M., Schmitt, W. W., and Bachmann, L. (1972): Freeze etching of cells without cryoprotectants. *J. Cell Biol.,* 53:116–126.
10. Rash, J. E. (1974): Discussion. In: *Exploratory Concepts in Muscular Dystrophy II. Excerpta Medica Int. Congr. Ser. No.* 333:243–244.
11. Streit, P., Akert, K., Sandri, C., Livingston, R. B., and Moor, H. (1972): Dynamic ultrastructure of presynaptic membranes at nerve terminals in the spinal cord of rats. Anesthetized and unanesthetized preparations compared. *Brain Res.,* 48:11–26.

Freeze Fracture: Methods, Artifacts, and Interpretations, edited by J. E. Rash and C. S. Hudson. Raven Press, New York © 1979.

A Simple Guide to the Evaluation of the Quality of a Freeze-Fracture Replica

L. Andrew Staehelin

Department of Molecular, Cellular, and Developmental Biology, University of Colorado, Boulder, Colorado 80309

Every primary user of the freeze-fracture technique is regularly confronted with the question: How does one recognize a "good" replica? Two approaches can be taken to define what is "good": one based on the exclusion principle, where one defines what should be avoided, and the other on "positive indicators." The latter approach forms the basis for this communication.

In contrast to the deceptive simplicity of the opening question, providing a substantial answer is a difficult and complex task, since many of the criteria used are based on general experience rather than on proven facts. Thus, any comprehensive answer has to include both factual information and a subjective interpretation of numerous personal observations as well as the published results of others. Because of space limitations no attempt has been made to cover this topic in a comprehensive or even systematic manner. Instead the ideas have been developed in the form of a short practical guide for the evaluation of the quality of a freeze-fracture replica. It should be stressed, however, that long-term success in freeze fracturing depends not only on the researcher's ability to recognize a good replica, but even more so on the ability to determine what remedial action must be taken if a nonoptimal replica is obtained. This type of troubleshooting requires, above all, an in-depth knowledge of the main physical parameters governing each of the steps of the procedure. A good source of such information is the book *Freeze-Etching, Techniques and Applications,* edited by Benedetti and Favard (1).

POSITIVE PARAMETERS OF A "GOOD" FREEZE-FRACTURE REPLICA

The following is a brief list of positive indicators that can be exploited to obtain a quick evaluation of the overall quality of a freeze-fracture replica and of the sample that has been replicated:

(a) The heavy metal shadowing material exhibits a very fine grain (point-to-point resolution better than 30 Å), that is stable under the electron beam.

(b) The replica possesses a good, crisp contrast; it has a "sparkle" when viewed in the electron microscope.

(c) Complementary-type fracture faces (P- and E-faces of the same type of membrane) exhibit a good "fit." In particular, the number of pits on one face should approximate the number of particles on the other. This rule does not apply when a significant amount of plastic deformation occurs during the fracturing process.

(d) Cells and organelles should have a turgid appearance (i.e., their membranes should have relatively smooth contours, not wavy ones).

(e) Each membrane system should be recognizable by its unique three-dimensional form and by its unique supramolecular architecture.

(f) Particles seen on fracture faces of membranes should be evenly distributed or show a defined mosaic structure; a diffuse, patchy distribution of particles is usually only seen on damaged membranes.

(g) The frozen medium around the sample in a nonetched preparation should exhibit a fine granular texture and fracture marks; a lumpy surface often signals water vapor contamination that may give rise to artificial intramembrane particles.

The freeze-fractured chloroplast membranes of spinach shown in Fig. 1 provide illustrations of points a and b of the preceding list. Although this micrograph has been enlarged to ×184,000 the granularity of the shadowing material is hardly seen. In addition, the replica exhibits a good contrast as well as a full range of gray tones. The high resolution of the micrograph is further illustrated by the clarity of definition of each of the 60 to 80 Å particles on the P-face (PFs), and the appearance of a distinct substructure on many of the 120 to 160 Å E-face (EFs) particles. Note also the presence of numerous large pits *(arrowheads)* on the P-face left behind by the large EFs particles. The smaller pits on the E-face of Fig. 1 are not seen because of the relatively high shadow angle.

The criterion of matching complementary membrane faces (point c of list) is probably the most sensitive and the most straight-forward method for judging the quality of a freeze-fracture replica. Figure 2 demonstrates what might be considered a good "fit" between complementary-type fracture faces of the plasma membrane of two forming daughter cells of the bacterium *Streptococcus faecalis*. In this micrograph the evenly distributed, small (40 to 70 Å) P-face particles are excluded from smooth, round areas, that arise owing to lipid phase transitions caused by cooling of the cells from 37 to 20°C. The resulting pattern is also reflected in the distribution of E-face pits, whose density approximates the density of P-face particles. As highlighted by Figs. 1 and 2, the clear visualization of particle-matching pits depends both on the overall quality of the replica and on the local shadowing angle (the optimal shadowing angle for visualizing pits may vary for different kinds of pits).

Points d, e, and f of the check list relate to structural membrane parameters that can be affected by nonoptimal pretreatment and/or freezing conditions.

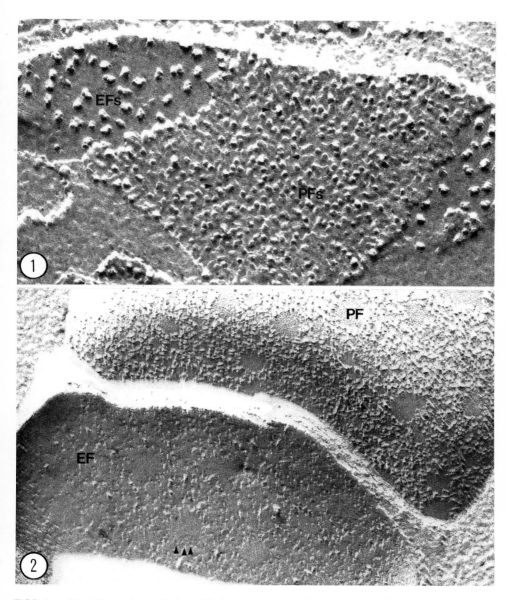

FIGS. 1 and 2. 1: Example of a high quality freeze-fracture replica of spinach chloroplast membranes. EFs and PFs stand for E- and P-faces of stacked *(grana)* membrane regions. Note the fine granularity and good contrast of the replica, and the presence of numerous large pits *(arrowheads)* on the PFs face produced by the tearing away of the large EFs particles. ×184,000. **2:** Freeze-fracture image of dividing cells of the bacterium *Streptococcus faecalis* selected to illustrate the matching of complementary-type fracture faces in a good replica. It can be seen that the pattern of small pits and smooth, round patches on the E-face closely correspond to the pattern of small particles (40 to 70 Å in diameter) and smooth, round patches on the complementary-type P-face. The density of E-face pits *(arrowheads)* is very similar to the density of P-face particles. ×156,000.

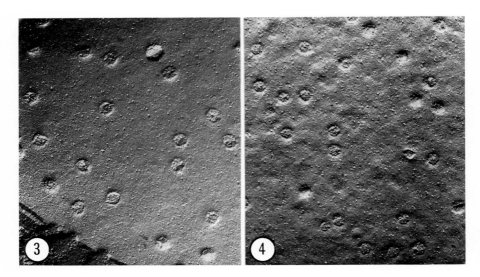

FIGS. 3 and 4. Freeze-fractured nuclear envelope membranes of dissociated chick embryo liver cells. Although the membrane of the well-preserved cell (Fig. 3) has very smooth contours, the envelope of the nonoptimally preserved cell (Fig. 4) appears distinctly uneven. Fig. 3: ×41,300. Fig. 4: ×36,000.

In well-preserved cells, all membrane-bound organelles have a turgid appearance and the membranes themselves exhibit relatively smooth contours (Fig. 3). Cells having undergone shrinkage owing to osmotic changes in the surrounding medium or to glutaraldehyde fixation etc., often possess membranes that exhibit a general waviness (Fig. 4). At a more refined level one finds that in well-preserved cells every portion of the cytoplasm and every portion of each membrane system has the appearance of being precisely organized. As a result, each membrane system may be recognized by its unique shape, unique disposition within the cell, and unique supramolecular architecture (Fig. 5). Damage of a cell incurred for example, during glycerination, even after glutaraldehyde prefixation, often manifests itself by a loss of structural differentiation between different membrane systems and by a general increase in membrane vesiculation (Fig. 6). The changes seem to occur gradually and can vary from quite subtle to fairly pronounced.

Well-preserved membranes usually demonstrate an even distribution or a well-defined, mosaic-like organization of intramembrane particles (point f of list). This organization often gives way to a patchy distribution of particles and to particle-free membrane regions if the cell is subjected to stress conditions (mechanical, chemical, osmotic, or temperature stress) during pretreatment. However, as discussed in other chapters of this volume, although most particle-free, blister-like membrane structures appear to represent artifacts, some may be genuine features of healthy membranes.

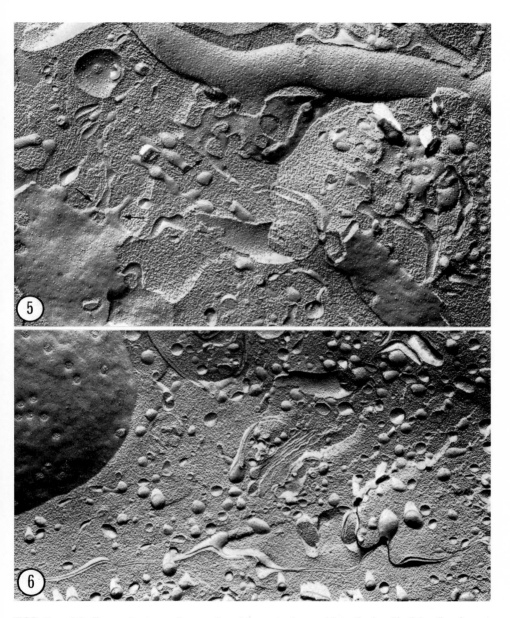

FIGS. 5 and 6. Freeze-fracture micrographs of the cytoplasm of intestinal epithelial cells of a rat. In the well-preserved cell (Fig. 5) each membrane system has a unique architecture. The transitions from sheet-like regions of the rough endoplasmic reticulum (ER) to the more tubular smooth ER regions *(arrows)* are distinct, indicating that they are under precise cellular control. In contrast, the poorly preserved cell (Fig. 6) shows much less differentiation between the different membrane systems, and most of the ER membranes have become vesicular in nature. Fig. 5: ×35,900. Fig. 6: ×25,000.

The last point on the check list (point g) pertains to changes in the appearance of membrane faces caused by water vapor condensation. In physical terms, condensation of water vapor onto a frozen specimen occurs if its saturation vapor pressure at the specimen temperature is smaller than the partial pressure

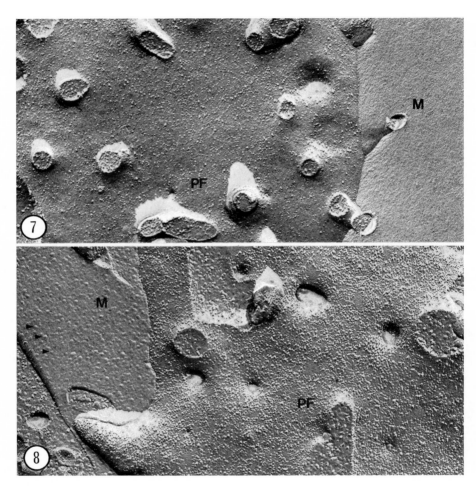

FIGS. 7 and 8. Freeze-fracture images of dissociated chick embryo liver cells. In the uncontaminated specimen (Fig. 7) the frozen medium (M) surrounding the cell appears amorphous and shows fine fracture marks. Similarly the P-face of the plasma membrane exhibits a fine granular background containing pits and several clear categories of sometimes closely spaced but not overlapping particles. In the specimen contaminated with water vapor (Fig. 8), on the other hand, the surface of the frozen medium (M) is covered with rounded bumps *(arrowheads)*, and the P-fracture face of the membrane is covered with very numerous irregular particles, some of which appear to be partly fused. No pits can be recognized and particle-free membrane regions (not shown) possess a coarse granular to lumpy surface texture as well. Figs. 7 and 8: ×43,000.

of water vapor in the vacuum chamber. Since the saturation vapor pressure of water at $-110°C$ is 10^{-6} torr, any specimen processed below $-110°C$ in a freeze-etch apparatus operated at a vacuum of 10^{-6} torr is subject to more or less significant water vapor condensation, i.e., contamination. As pointed out by Staehelin and Bertaud (2) the appearance of the frozen medium (usually glycerol–water) surrounding the sample can be used to evaluate the presence of water vapor contamination of a freeze-fracture specimen, since the water molecules condense preferentially onto the frozen water phase. This causes a buildup of condensation on the ice crystals, thereby increasing their height with respect to the surrounding glycerol hypoeutectic phase. Figure 7 depicts a freeze-fracture micrograph of a dissociated chick embryo liver cell from an uncontaminated specimen. The medium surrounding the cell has a relatively smooth texture, modified only by fine fracture marks. In contrast, the frozen medium surrounding the cell shown in Fig. 8 has a lumpy appearance and lacks fracture marks, thus indicating that it has been contaminated by water vapor. A comparison of the membrane fracture faces of Figs. 7 and 8 reveals further that the water vapor contamination has also increased the apparent number of intramembrane particles quite significantly.

In conclusion, it is hoped that by providing simple directions for the evaluation of the quality of freeze-fracture replicas this brief communication will contribute to the overall quality of future studies in which this technique is used.

ACKNOWLEDGMENTS

The excellent technical assistance of Marcia De Wit is gratefully acknowledged. Supported by grant 18639 of the National Institute of General Medical Sciences.

REFERENCES

1. Benedetti, E. L., and Favard, P. (1973): Freeze-etching, techniques and applications. Société Française de Microscopie Electronique, Paris.
2. Staehelin, L. A., and Bertaud, W. S. (1971): Temperature and contamination dependent freeze-etch images of frozen water and glycerol solutions. *J. Ultrastruct. Res.*, 37:146–168.

Freeze Fracture: Methods, Artifacts, and Interpretations, edited by J. E. Rash and C. S. Hudson. Raven Press, New York © 1979.

Artifacts and Defects of Preparation in Freeze-Etch Technique

S. Böhler

Balzers AG, Balzers, Fürstentum Liechtenstein

Originally, freeze-etch technique was regarded as an electron microscopical specimen preparation method, enabling the examination of biological material in the electron microscope, nature true and free of artifacts. Past experience has shown, however, that this was not the case. The causes of changes in the specimens during preparation arising from the freeze-etch technique are various. They can be found in the characteristics of the preparation method or of the specimen itself which is being investigated, or they can be caused by careless preparation or defective or inadequate equipment.

As the introduction of artifacts or of faulty preparation is possible in each of the six preparation steps (1), they can be classified as follows:

(a) Artifacts caused by antifreeze agents.
(b) Changes in the specimen during freezing.
(c) Artifacts caused by cutting or fracturing of the frozen specimen.
(d) Sublimation and condensation artifacts.
(e) Artifacts and faults during coating.
(f) Insufficient cleaning of the specimen replica.

Although each of these areas is discussed in greater detail in other chapters in this volume, the following description should place the problems in perspective.

ARTIFACTS CAUSED BY ANTIFREEZE AGENTS

It is widely known that, for example, mitochondria are subject to morphological changes owing to the effect of glycerol, which is used as an antifreeze agent to prevent the formation of intracellular ice crystals. Mitochondria swell up strongly while absorbing the glycerol. Unfortunately, when specimens with a high water content are to be frozen by conventional means, that is, liquid Freon, liquid propane, or subcooled nitrogen, they must be frozen in an antifreeze agent. The only way to avoid swelling of mitochondria owing to the effect of glycerol is to employ a slight chemical fixation of the specimen with, for example, 2% glutaraldehyde before adding the antifreeze agent.

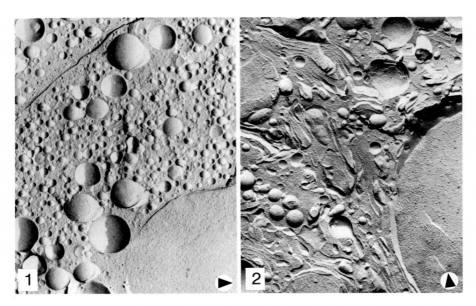

FIGS. 1 and 2. 1: Liver cell incubated in 30% glycerol without chemical fixation: endoplasmatic reticulum (ER) transformed in vesicular system. (M. A. Spycher). **2:** Liver cell prefixed with 2% glutaraldehyde before incubating in 30% glycerol; ER shows original laminar system. (M. A. Spycher).

Another example of altered ultrastructure of cells or cell organelles owing to the effect of glycerol is the transformation of the original laminar system of the endoplasmatic reticulum into a vesicular system (Fig. 1). This structural change can also be prevented by a mild chemical fixation of the specimen (Fig. 2) or by ultrarapid freezing (3). Freezing of specimens without an antifreeze agent and subsequently without chemical fixation is only possible with advanced freezing methods such as high-pressure freezing, propane jet freezing, or spray freezing. These methods or brief fixation in 2% glutaraldehyde may be useful in minimizing cryoprotectant-induced artifacts (see C. F. Hudson, et al. and G. G. Maul, *this volume;* but see also cautionary notes by E. D. Hay and D. L. Hasty; E. Shelton; and D. E. Chandler, *this volume*).

CHANGES IN THE SPECIMEN DURING FREEZING

When living cells are frozen without special specimen preparation, the decisive factor for their survival is the freezing speed. The lethal crystallization of the cell water can be prevented both by a very low (0.01°C/sec) or a very high freezing rate (1000°C/sec), whereas a medium freezing speed from 1 to 10°C/sec unquestionably destroys the cell because of the growing ice crystals (Fig. 3).

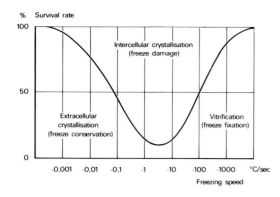

FIG. 3. Survival rate of yeast cells at different freezing speeds. (From Moor, ref. 2, with permission.)

A cell which is frozen at a very low speed can be kept alive. Its structure, however, will be changed by extracellular crystallization of the cell water and therefore cannot be used for morphological investigation. It becomes dehydrated by the surrounding ice if the cell is frozen extremely slowly. The water thus leaving the cell crystallizes on its outer membrane. Needle-shaped crystals form, surrounding the cell. The extraction of the water enables vitrification of the cell, but leads to a considerable loss in volume hence completely deforming the cell (2).

A cell frozen at an average speed of approximately 100°C/sec does not change

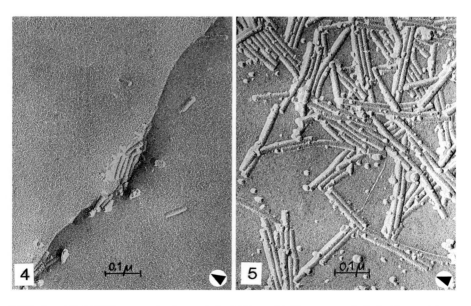

FIGS. 4 and 5. 4: TMV suspension frozen in liquid Freon 22. Virus concentrated at the periphery of the ice crystals (M. Müller). **5:** TMV suspension frozen in propane jet. Specimen thickness 15 μm. Viruses uniformly distributed (M. Müller).

its exterior shape. However, because of intracellular crystallization of the cell water, the cytoplasmatic membrane systems are completely destroyed (2). The higher the free water content of the cell, the greater is the danger of ice crystal formation. With the normal cryofixation applied in the freeze-etch technique using liquid Freon or similar substances, the ice crystal formation can be prevented or at least kept in reasonable limits only by using antifreeze agents. However, as we already know, these antifreeze agents can also lead to preparation artifacts.

Figure 4 shows another example of a freezing artifact. When freezing suspensions in liquid Freon, dissociation of dissolved or colloidally distributed materials frequently occurs because the freezing rates are too low. This causes the materials to suffer artifact displacement, and as a result they accumulate in high concentration at the periphery of the ice crystals. Such dissociation of artifacts can be prevented only by increasing the freezing speed. As Fig. 5 shows, the viruses of a tobacco mosaic virus (TMV) suspension, frozen with the propane jet freezing technique are not dissociated but remain uniformly distributed (3).

ARTIFACTS OBTAINED BY CUTTING OR FRACTURING OF THE FROZEN SPECIMEN

If a frozen specimen is cut with a deep-cooled knife, a fracture surface is achieved rather than a smooth cutting surface; the reason is that the specimen, mainly comprising ice, is very brittle and the material splits off when cut. It is rather a fracturing than a cutting process since the knife edge normally does not touch the created fracture surface. Should this happen, knife marks can be seen on the "cutting surface." The exposed specimen structures will then be totally indistinct. Knife marks can be reduced by increasing the cutting speed and the rate of knife advance.

If a frozen tissue specimen is fractured instead of cut, material sometimes splits from the specimen fracture surface. The piece of tissue which splits off bends upward from the fracture surface, but stays connected to it. Such a projecting flake of material is also coated during replication and contrasts in the freeze-etch micrograph as a dark spot (Fig. 6). Such splitting off of material cannot be avoided when a frozen piece of tissue is fractured.

Another very unpleasant freeze-etch artifact is the plastic deformation of specimen material. In freeze-etching technique, materials sometimes have to be prepared which are plastically deformable even at very low temperature, as for example polyphosphate or polyhydrobutyrate bodies in bacterias (4). If such a plastically deformable particle is touched by the microtome knife during cutting, it is pulled out of the specimen surface. A horn-shaped formation arises projecting far out of the specimen fracture surface (Fig. 7). These deformation artifacts occur, of course, not only during cutting but also during fracturing of a frozen specimen. They cannot, unfortunately, be avoided and are visible in frozen specimens which are cut or fractured even at liquid nitrogen temperature.

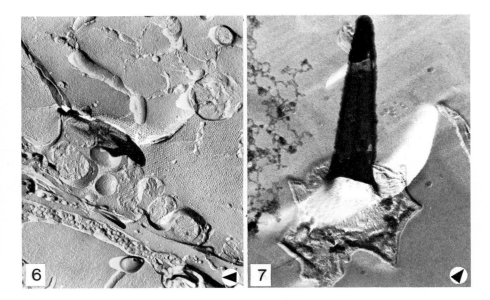

FIGS. 6 and 7. 6: Splitting of specimen material during fracturing. **7:** Hyphomicrobium B 522. Plastic deformation of specimen granule. Hyphomicrobium B 522 is a special type of this microorganism.

SUBLIMATION AND CONDENSATION ARTIFACTS

For a better understanding of the relation between condensation and sublimation it is necessary to examine the diagram of the saturation vapor pressure curve for water, which is the dominating component of the residual gas in the working chamber under normal vacuum conditions.

Figure 8 shows that at a specimen temperature of $-100°C$ the saturation vapor pressure will be 1×10^{-5} torr (point 1). If the water vapor partial pressure above the surface of the specimen which is kept at $-100°C$ in the vacuum

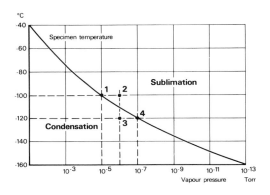

FIG. 8. Saturation vapor pressure of water.

chamber is also 1×10^{-5} torr, sublimation and condensation are in balance. In order to avoid as far as possible condensation of water vapor on the specimen surface, the water vapor partial pressure must be lowered. This means the vacuum has to be increased to 1×10^{-6} torr or even better while keeping the specimen temperature at $-100°C$. Thus, as shown, we are approaching the region where sublimation takes place (point 2). If, for instance, the specimen temperature is lowered to $-120°C$, a vacuum of 1×10^{-6} torr would still keep us in the range of condensation (point 3). At this temperature the saturation vapor pressure at which sublimation and condensation is in balance would be reached only at approximately 1×10^{-7} torr (point 4). If one wants to operate at a specimen temperature of $-120°C$ and at the same time wants to avoid contamination by water vapor condensation, the vacuum must be better than 1×10^{-7} torr, which means that it must be at least in the 10^{-8} torr range.

As the first example of the group "sublimation and condensation artifacts" the glycerol etch patterns can be considered. During freezing a drop of water ice crystals are formed; the lower the freezing speed, the larger are the crystals. Only pure water crystallizes. Dissolved or dispersed materials dissociate, are displaced by the growing ice crystals and form a network throughout the whole specimen. All the ice crystals of a frozen glycerol–water mixture are therefore surrounded by dissociated glycerol. If such a frozen specimen drop is fractured and "etched," the pure ice sublimes much faster than the glycerol. The latter remains as a web between the ice crystals and thus forms the typical glycerol etch patterns.

During "etching" of a frozen specimen the sublimation rate for ice increases rapidly with increasing specimen temperature (Fig. 9). Although it is only 15 Å/sec at $-100°C$, it is already 500 Å/sec at $-80°C$ and 10,000 Å/sec at $-60°C$. Therefore, if a specimen is etched at too high a temperature owing to a defect or insufficently accurate temperature measuring equipment, very deep etch-cavities arise, as shown in Fig. 10. These etch cavities are all decorated

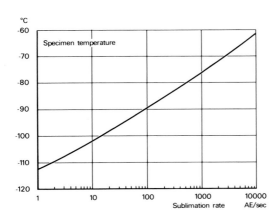

FIG. 9. Sublimation rate of ice under appropriate vacuum conditions.

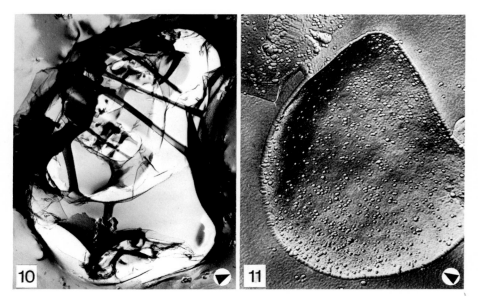

FIGS. 10 and 11. 10: 10% water-glycerol solution etched at −80°C for 10 min. **11:** Heavy contamination of specimen by water vapor condensation.

with remaining glycerol structures. The subliming process of the ice only stops if the whole remaining specimen surface is covered with a glycerol layer.

"Etching" of a frozen specimen produces water vapor as the result of the sublimation of ice, thus considerably increasing the water vapor partial pressure which is already high, particularly in the specimen area. Parts of this water vapor can be pumped off but other parts condense again on cold surfaces such as the specimen surface. Because of this phenomenon, the surface is contaminated differently, a fact that shows a very specific interaction between substrate and condensate (5).

The condensation of water on the pure ice surface between the cells of a cell suspension is homogenous. If the specimen is only slightly contaminated, as it may occur because of careless covering of the fracture surface during "etching," the formation of particles is roughly homogenous all across the ice surface.

If the specimen temperature is too low, that is, if the preparation work is carried out at condensation conditions for water vapor, the specimen is heavily contaminated as shown in Fig. 11. In this case a homogenous contamination film grows up on the pure ice surface between the cells, whereas on the membrane surfaces ice crystals of various sizes are formed that sometimes can be clearly recognized as cubic-shaped crystals. Sharp contours, such as membrane fracture edges can be completely blurred because of heavy contamination by condensed water molecules.

ARTIFACTS AND FAULTS DURING COATING

Contrast and resolution of a freeze-etch micrograph can be increased by suitable shadowing of the specimen. For this shadow casting, normally a platinum–carbon mixture is deposited at a film thickness of approximately 20 Å.

If the proportion of platinum in the shadowing film is too low because of a faulty evaporation, or the quantity of platinum used for evaporation is too small, or if the film thickness selected is too thin, the freeze-etch micrograph results in poor contrast, and fine structure details can only be recognized with difficulty or not at all (Fig. 12).

Another problem, encountered during coating is damage of the specimen surface owing to high thermal load. This thermal load is composed of (a) the radiation heat of the evaporation material, that is, from the evaporation source, (b) the condensation heat of the deposited coating film, and, (c) if electron-beam guns are used, the heat that is produced by the bombardment of the specimen surface with charged particles. Technically well-designed electron-beam guns, however, are provided with highly efficient deflecting systems for charged particles so that bombardment of the specimen surface is prevented. Moreover, the condensation heat is relatively low owing to the low condensation rates so that in practice only the radiation heat from the evaporation source has to be considered. Because the thermal conductivity in biological material is extremely low, even with deep-cooled samples, a thermal load too high leads

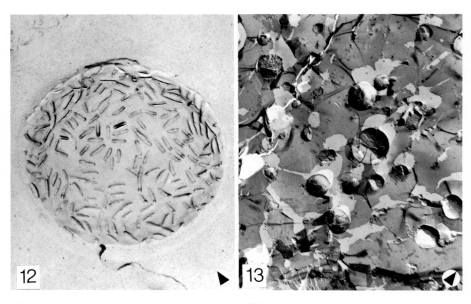

FIGS. 12 and 13. 12: Extracellular fracture face (EF) of yeast plasma membrane shadowed with 5-Å platinum–carbon. **13:** Shifting of carbon backing film on platinum–carbon shadowing film.

to artificial changes of the specimen surface. It melts owing to the effect of the radiation heat sometimes causing the specimen structures to be more or less destroyed. Heat radiation damage may be minimized by utilizing an automatic shutter (see M. H. Ellisman and L. A. Staehelin, *this volume*).

A rather strange phenomenon in connection with the specimen coating is the separation of the shadowing film from the carbon replica. When making a replica of a freeze-etched specimen. it is first shadowed with platinum–carbon and then coated with pure carbon. If the two layers are, for example, for technical reasons, not evaporated immediately one after the other, and if the vacuum conditions are not good enough, a thin layer of water vapor will be formed on the platinum–carbon film which freezes immediately because of the low specimen temperature. In this case, the carbon film is not deposited on the platinum–carbon film but on a thin layer of ice. When the replica is being separated from the specimen, the ice layer between the two coating films thaws and the carbon backing film may shift on the platinum–carbon layer (Fig. 13).

CONTAMINATION CAUSED BY INSUFFICIENT CLEANING OF THE SPECIMEN REPLICA

Specimen contamination can be divided into two groups. The first comprises contamination caused by water vapor or hydrocarbon condensation on the cold specimen surface in the vacuum chamber, the second consists of material other than condensate from the vacuum. This second group, which will be discussed in the last section, includes:

(a) Contamination by ice crystals and dirt from the liquid nitrogen which may fall on the specimen surface when the specimen is not cut or fractured in a vacuum chamber but under liquid nitrogen.

(b) Contamination of the replica with undigested specimen material.

(c) Contamination of the replica with crystals and dirt from the cleaning solvents.

General directions for cleaning a specimen replica cannot be given. A suitable solvent can sometimes only be found by carrying out a number of tests and it must be adapted to the various specimens. For example, sodium hypochlorite alone or preceded by 40% chromic acid followed by a short washing is normally suitable for animal tissues. If the wrong solvent is used or the cleaning time is too short, residues of the specimen remain on the replica (Fig. 14). These contaminated parts appear as diffuse, dark patches on the freeze-etch micrograph.

If tissue with a high fat content or specimens with pure fat inclusions are freeze etched, special care has to be taken in the cleaning of the specimen replica from the residues. After separating the specimen from the replica it has to be treated with an agent for dissolving fat. In this case it is vital that the final step in the cleaning process is the use of a cleaning agent for dissolving fat, such as for example acetone or methanol.

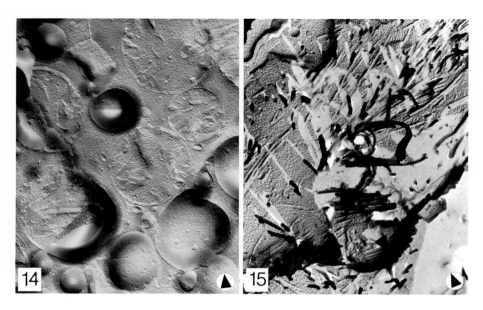

FIGS. 14 and 15. 14: Tissue remnants on replica after too short a cleaning time. **15:** Collagen fibrils not dissolved because of the use of the wrong cleaning fluid.

If a specimen replica of a collagen-containing tissue is cleaned with chromic acid and sodium hypochlorite, as it is usual for animal tissue, the collagen remains as dark fibers (Fig. 15). The main reason for this is that the material is difficult to remove from the narrow sheath of the fibrils protruding out of the specimen surface and also the collagen fibrils do not dissolve well with chromic acid. In this case, it is advisable to first clean the specimen replica with chromic acid, then with sodium hypochlorite, and afterward to immerse it for some time in diluted potassium hydroxide, since this dissolves the collagen fibrils. Whatever is used for cleaning of the replica, the final step of this procedure is always a careful rinsing of the replica in double-distilled water.

Most of the problems concerning artifacts that can arise in freeze-etch technique can be avoided by careful work and suitable equipment. The few inevitable ones, however, lose their dangers if they have been recognized and suitably considered.

REFERENCES

1. Böhler, S. (1975): Artefacts and specimen preparation faults in freeze-etch technology. Balzers Publication DN 6573, Fürstentum Liechtenstein.
2. Moor, H. (1964): Die Gefrier-Fixation lebender Zellen und ihre Anwendung in der Elektronenmikroskopie. *Z. Zellforsch.*, 62:546–580.
3. Müller, M., Meister, N., and Moor, H. Freezing in a propane jet and its application in freeze-fracturing. *J. Cell Microscop.* Manuscript submitted for publication.

4. Van Gool, A. P., Lambert, R., and Laudelout, H. (1969): The fine structure of frozen-etched Nitrobacter cells. *Arch. Mikrobiol.,* 69:281–293.
5. Staehelin, L. A., and Bertaud, W. S. (1971): Temperature contamination dependent freeze-etch images of frozen water and glycerol solutions. *J. Ultrastruct. Res.,* 37:146–168.

Freeze Fracture: Methods, Artifacts, and Interpretations, edited by J. E. Rash and C. S. Hudson. Raven Press. New York © 1979.

Artifacts Associated with the Deep-Etching Technique

Kenneth R. Miller

The Biological Laboratories, Harvard University, Cambridge, Massachusetts 02138

Although the greater part of "freeze-etching" literature has concerned itself with the freeze-fracture method of specimen preparation in which a fractured specimen is immediately used to form a metal replica, there is another way to use the freeze-etching technique to study biological material. This other method is called deep-etching and involves the sublimination of frozen solute to expose macromolecules or membrane surfaces just below the fracture plane. This technique was used by Pinto da Silva and Branton (5) to confirm that the true surface of the erythrocyte membrane was indeed different from the membrane faces exposed by freeze-fracturing. The principal advantage of deep-etching, in addition to providing two new views of the membrane being studied, is that the inner and outer surfaces of the membrane thus revealed are true surfaces. Such true surfaces can be studied directly by the addition of probes such as ferritin labels, and are free of the distortions of the fracturing process.

The principal artifacts of the deep-etching technique are those of freezing itself. The growth of ice crystals in specimens frozen for deep-etching was directly observed by Staehelin and Bertaud (6) who demonstrated that glycerol and water solutions were rapidly partitioned into two phases during the freezing process. They suggested that these two phases were formed by the growth of large ice crystals in certain regions of the specimen. Glycerol is generally not included in specimens intended for deep-etching because it is not volatile. However, other materials intended for study can be affected in the same way by the freezing process. Figures 1 and 2 show the effect of conventional freezing in liquid freon on a preparation of histone 1 depleted chicken erythrocyte chromatin (prepared in 5 mM phosphate buffer, pH 7.0). Deep-etching of the frozen sample has been used to expose nucleosomes (Fig. 2). The direct visualization of individual nucleosomes indicates the practicality of using the deep-etching technique as a method for analyzing higher-order structures of chromatin. However, the formation of ice crystals has squeezed the chromatin into highly compacted regions where the overall structure of the material has been severely altered. The use of the deep-etching technique to solve the chromatin problem, therefore, must await the application of methods which will slow or prevent the growth of ice crystals.

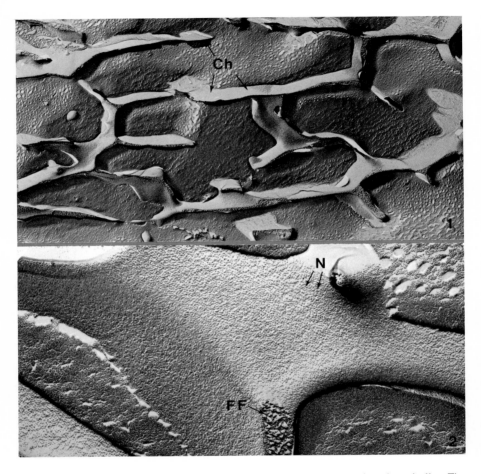

FIGS. 1 and 2. 1: Suspension of purified chromatin, frozen in 5 mM phosphate buffer. The sample has been fractured and etched for 5 min at −100°C (conditions identical to those for all subsequent micrographs). Ice crystals have forced the chromatin (Ch) into condensed areas where its structure has been greatly distorted. ×12,000. **2:** Higher magnification of the chromatin preparation shown in Fig. 1. Individual nucleosomes (N) can clearly be seen at the surface of the chromatin mass. The region where the chromatin mass was broken by the fracture plane (FF) is also indicated. Although individual nucleosomes can be visualized by this method, ice crystal formation distorts their secondary structure. ×53,000.

The problem of ice crystal formation is less severe in the case of biological membranes, since an intact membrane is less susceptible to damage caused by crystal growth than the delicate meshwork of chromatin fibers. As a model system for the studies reported here, I have chosen the photosynthetic membrane of the higher-plant chloroplast. The deep-etching technique has been used to great advantage in our laboratory in studies of this membrane (1–4). The photosynthetic membrane, or thylakoid, also has the advantage that its membrane surfaces are each covered with a distinct and characteristic distribution of particles, so that artifacts of the deep-etching process can be easily recognized against the details of the membrane surface.

For most of our published studies, the thylakoid membrane has been frozen in dilute (less than 10 mM) buffers, and the true membrane surfaces revealed by etching after freeze fracturing. Even at these low buffer concentrations, the boundaries of ice crystals show clearly on the etched membrane surfaces, as seen on Fig. 3. Here the boundary of eutectic between two ice crystals crosses the outer surface of a photosynthetic membrane. Although the boundary interferes

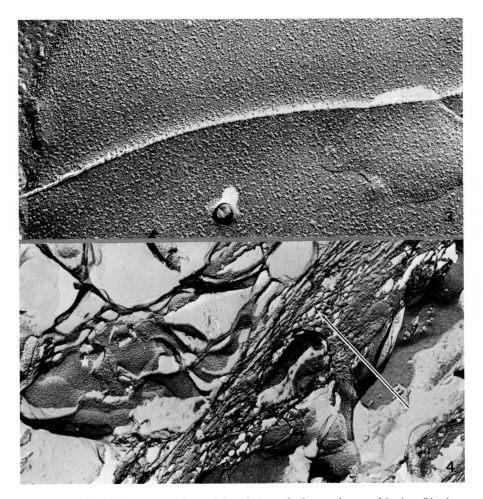

FIGS. 3 and 4. 3: The outer surface of the photosynthetic membrane of barley *(Hordeum vulgare)* exposed by deep etching. This sample was frozen in 5 mM NaCl. Although most of the membrane surface is clear and unobscured, a boundary between two ice crystals runs across the membrane from left to right. ×56,000. **4:** Barley thylakoid membranes frozen in 200 mM NaCl. Although membrane surfaces can be observed on the left side of the micrograph, at the right side a large mass of nonetchable material *(arrow)* obscures detail. ×33,000.

FIGS. 5–7. The inner surface of the thylakoid membrane, as observed in various buffer concentrations. Fig. 5: 5 mM; Fig. 6: 200 mM and Fig. 7: 500 mM. Although artifacts increase at the boundary regions between adjacent ice crystals, regions of the membrane surface where fine details are observable are still present, even in 500 mM buffer. ×96,000.

with direct observation of the membrane, its small size makes it more an esthetic problem than a scientific one in the study of membrane surfaces.

However, as the concentration of dissolved material increases, the amount of material forced into the boundary eutectic by the freezing process also increases. Figure 4 shows a sample of thylakoid membranes frozen in 200 mM NaCl. Although some membrane surfaces are clearly visible, a large region of nonvolatile material contaminates the remainder of the micrograph. Such regions are common in membrane samples frozen in high-solute concentrations. A prime reason for preparing membranes in buffers of lower-solute concentration, therefore, is to reduce the extent and severity of this artifact.

When experimental conditions make this problem unavoidable, it is nevertheless still possible to make detailed observations of membrane surface structure if the investigator is able to visualize membrane surfaces between the boundary regions. This is shown in Figs. 5 to 7. The inner surface of the photosynthetic membrane is characterized by a series of unique particles, most often showing four subunits (Fig. 5). The ability to resolve this substructure is a sensitive measure of the extent to which uncontaminated membrane surfaces can be accurately observed on a deep-etch replica. Figures 6 and 7, prepared from samples frozen in 200 mM and 500 mM NaCl, respectively, show that accurate determination of membrane surface structure is still possible in samples frozen at high solute concentrations, if one is willing to deal with the problems presented by ice crystal boundaries. Incidentally, the outer surface of the membrane can also be observed with clarity in these preparations.

Deep-etching, therefore, is a sensitive and useful method of determining the surface structure of biological material. Its sensitivity is limited by the usual problems of vacuum contamination and shadowing methodology, and most seriously, by the problems associated with ice crystal formation. The possibility that many of these problems can be solved by freezing biological samples at higher rates holds exciting prospects for future studies with the deep-etching technique.

ACKNOWLEDGMENTS

This work was supported by NIH grant GM-24078-02. I would like to thank Ms Gayle Miller for expert technical assistance and Dr. John Wooley for providing chromatin preparations.

REFERENCES

1. Miller, K. R. (1976): A particle spanning the photosynthetic membrane. *J. Ultrastruct. Res.,* 54:159–167.
2. Miller, K. R. (1978): Structural organization in the photosynthetic membrane. In: *Chloroplast Development,* edited by G. Akoyunoglou and J. H. Argyroudi-Akoyunoglou, pp. 17–30. Elsevier/North-Holland Biomedical Press, Amsterdam.
3. Miller, K. R., Miller, G. J., and McIntyre, K. R. (1976): The light-harvesting chlorophyll–

protein complex of photosystem II. Its location in the photosynthetic membrane. *J. Cell Biol.,* 71:624–638.

4. Miller, K. R., and Staehelin, L. A. (1976): Analysis of the thylakoid outer surface. Coupling factor is limited to unstacked membrane regions. *J. Cell Biol.,* 68:30–47.

5. Pinto da Silva, P., and Branton, D. (1970): Membrane splitting in freeze-etching: Covalently bound ferritin as a membrane marker. *J. Cell Biol.,* 45:598–605.

6. Staehelin, L. A., and Bertaud, W. S. (1971): Temperature and contamination dependent freeze-etch images of frozen water and glycerol solutions. *J. Ultrastruct. Res.,* 37:146–168.

Freeze Fracture: Methods, Artifacts, and Interpretations, edited by J. E. Rash and C. S. Hudson. Raven Press, New York © 1979.

Temperature-Dependent Changes in Intramembrane Particle Distribution

Gerd G. Maul

The Wistar Institute of Anatomy and Biology, Philadelphia, Pennsylvania 19104

Freeze fracturing is a method for studying the structural organization of cell membranes. This technique allows the visualization of specific sites in the hydrophobic core of the membrane, the intramembranous particles. Membrane particles may change their distribution and appear partially aggregated when experimental conditions such as pH are changed (4). Moreover, interpretations of data are often based on this changing distribution, that is reversible (membrane fluidity, thermotropic lipid phase transitions, architectural rearrangement, temporal membrane specializations, etc.). However, any redistribution of intramembranous particles dependent on experimental preparation would render interpretation impossible.

Our experience with a variety of cell types has been that it is the nuclear and mitochondrial membranes, as opposed to the plasma membrane, that are the most sensitive to changes in intramembranous particle distribution. It will become evident from the description of the freezing process that, in the case of nuclear membranes, these changes are induced by shifts in temperature during initial tissue preparation.

In one experiment, we were attempting to determine what effect a reduced adenosine triphosphate (ATP) concentration in the rat liver had on the nuclear pore frequency. Ethionine was used to lower the ATP concentration. The first distinct effect was intramembranous particle segregation, that increased with length of ethionine exposure. We later learned that this was fortuitous. Figure 1 shows intramembranous particle distribution in isolated control nuclei. In contrast, the particle distribution 4 hr after the administration of ethionine shows an anastomosing network on the protoplasmic face (Fig. 2) and exoplasmic face (Fig. 3) of the inner and outer nuclear membranes. The maximal segregation of particles was observed 24 hr after ethionine administration (Fig. 3). In the center of most particle-free areas, larger and flatter membrane particles appeared (Fig. 4). These exciting results, initially attributed to the ethionine effect, became suspicious as soon as we repeated the experiments. Finding membrane particle segregation in our control nuclei, we were finally led to examine the preparatory sequence.

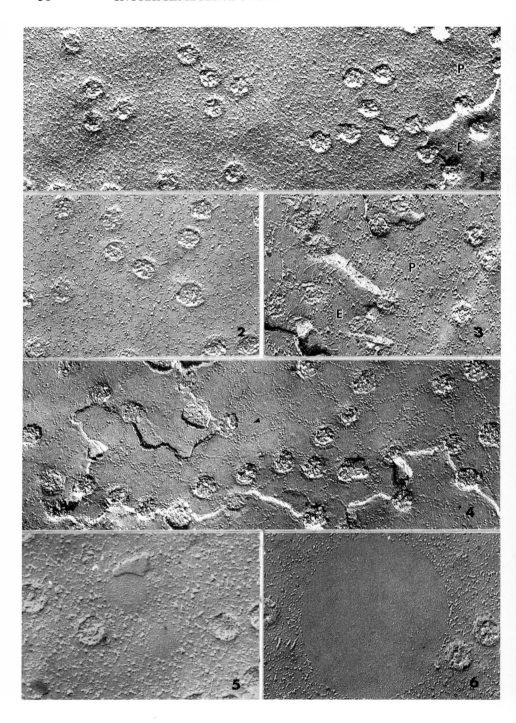

As is standard, we kept our unfixed nuclear suspension (0.25 M sucrose, 1 mM MgCl$_2$, 5 mM Hepes pH 7.4) on ice while slowly increasing the glycerol concentration to 20% over 20 min. After 30 min, the nuclear suspension was concentrated by centrifugation in a refrigerated centrifuge. After aspiration of the supernatant, freezing began. In preparation for this, the nuclear suspension was sucked up in a finely drawn pipet and a droplet placed on a gold specimen holder; this was then dipped into liquid nitrogen-cooled Freon 22. The pipet, the gold specimen holder, and the tweezers used to hold the latter were not cold. There was usually enough nuclear suspension in the pipet to make about 10 specimens. The length of time needed for freezing several specimens allowed most of the specimens to reach room temperature. Variable results within one time interval (different replicas had different degrees of particle segregation) were therefore attributed to the fact that some samples were frozen while they were still cold, whereas others had more time to equilibrate to room temperature. We, therefore, froze control nuclei in the cold room with the effect that their intramembranous particles were extremely segregated and none of the nuclear membrane was as "smooth" as after normal procedures; rather, they looked as "unhealthy" as after fixation in ice cold glutaraldehyde. Nuclei glycerinated and frozen at room temperature showed no particle segregation.

It is evident, then, that during specimen cooling intramembranous particles of unfixed isolated rat liver nuclei segregate into an anastomosing meshwork and that this arrangement is reversible, depending on the degree and length of time of specimen warming.

Nuclear membranes of intact cells did not show such particle distribution. In rat kidney nuclei, circular membrane particle-free areas were found (Fig. 5). These were assumed to be closing nuclear pores, since their distribution and center-to-center spacing correspond to that of the pore complexes (3). In other systems, large irregular membrane particle-free areas were found. Since the experiments of Wunderlich et al. (5) on the differential effects of temperature on the nuclear and plasma membranes of intact cells, it has been recognized that low temperatures can create these specialized membrane areas (Fig. 6). Wunderlich et al. (5) hypothesized ". . . that the changes observed in the nuclear membrane represent thermotropic lipid phase transitions and that such transitions either do not occur in the plasma membranes or are there constrained to very small regions." The work of Yu et al. (6) on membrane particle aggregation in erythrocyte membranes points to a constraining mechanism, that is, a

---

FIGS 1–6. 1: Apparently normal distribution of intramembranous particles over the outer and inner nuclear membranes of isolated rat liver nuclei. **2:** Segregated intramembranous particles on the protoplasmic face of a stretched outer nuclear membrane from isolated rat nuclei. **3:** Anastomosing network and particle-free areas, the results of intramembranous particle segregation, on the exoplasmic and protoplasmic faces of isolated rat liver nuclei. **4:** Maximally segregated intramembranous particles and a large particle in the center of a particle-free area. **5:** Regular, circular membrane particle-free areas in nuclei from rat kidney tissue. **6:** Large membrane particle-free areas in cells from chicken embryo explants. A few large flat particles are often found in these areas. Figs. 1–6: ×74,000.

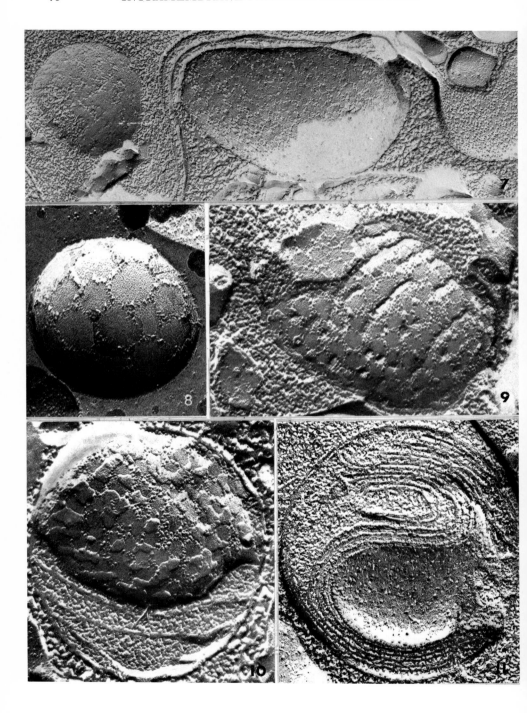

connection between the membranes and spectrin and actin. One possible explanation for the difference in particle segregation between isolated nuclei and nuclei in intact cells is the removal through isolation of certain components responsible for the apparently random distribution of particles in the normal state. Another explanation is that the tissue pieces had returned to room temperature during attachment to the copper disks, that is, before freezing. In the intact cell— despite glycerol—the redistribution may proceed faster.

Mitochondrial membranes also seem to show temperature-dependent membrane particle segregation (2). It shall be described here as a cautionary note with the hope of stimulating more controlled experimentation.

The normal appearance of mitochondrial membranes in rat kidneys is shown in Fig. 7. Large areas of membrane are easily obtained in this system. However, the fracture line often jumps between the two opposing membranes of the mitochondria, thereby exposing small areas not useful for membrane particle distribution analysis. The outer membrane pieces are often free from particles. In Fig. 8, a Rana cell culture held on ice during glycerination, one can see that particles have segregated in the center membrane into nearly single rows. In contrast, the nuclear membranes of the same cells show small circular particle-free areas but no linear array of particles. The E-face of the inner mitochondrial membrane is shown with the circular and slit connections of the cristae (Fig. 9). Although not as large, the areas free from particles here are reminiscent of those in isolated nuclei. Exposed crista membranes show the same pattern (Fig. 10). The question remains as to why some cisternal membranes show segregation of membrane particles and others do not (see structural question mark in Fig. 11).

One remedy for the temperature-dependent changes in membrane particle distribution is fixation with glutaraldehyde at room temperature or approximately 37°C. However, as described in other chapters of this volume, the technique may result in other artifacts and so may not be an alternative in all experiments. Moreover, fixation is a relatively slow process. Fast freezing as developed by Heuser et al. (1) may be the ultimate solution.

REFERENCES

1. Heuser, Y. E., Reese, T. S., and Landis, D. M. D. (1976): Preservation of synaptic structure by rapid freezing. *Cold Spring Harbor Symp. Quant. Biol.*, 40:17–24.
2. Höchli, M., and Hackenbrock, C. R. (1977): Thermotropic lateral translational motion of intramembrane particles in the inner mitochondrial membrane and its inhibition by artificial peripheral proteins. *J. Cell Biol.*, 72:278–291.

FIGS 7–11. 7: Normal mitochondrial membranes in rat kidneys. **8:** Exoplasmic face of the outer mitochondrial membrane of a Rana tissue culture cell. **9:** Protoplasmic face of the inner mitochondrial membrane of a Rana tissue culture cell. **10:** Exposed membranes of a mitochondrial cisterna. Note rows of particles and pits on inner and outer membranes. **11:** The "question mark" signifies our query as to why some cristae have segregated membrane particles and others do not. Figs. 7–11: ×76,000.

3. Maul, G. G., Price, J. W., and Lieberman, M. W. (1971): Formation and distribution of nuclear pore complexes in interphase. *J. Cell Biol.,* 51:405–418.
4. Pinto da Silva, P. (1972): Translational mobility of the membrane intercalated particles of human erythrocyte ghosts, pH dependent, reversible aggregation. *J. Cell Biol.,* 53:777–787.
5. Wunderlich, F., Hozel, D. F., Speth, V., and Fischer, H. (1974): Differential effects of temperature on the nuclear and plasma membranes of lymphoid cells. *Biochem. Biophys. Acta,* 373:34–43.
6. Yu, Y., Elgsaeter, A., and Branton, D. (1977): Intramembrane particle aggregation in erythrocyte membranes and band 3-lipid recombinants. *Prog. Clin. Biol. Res.,* 17:453–458.

Freeze Fracture: Methods, Artifacts, and Interpretations, edited by J. E. Rash and C. S. Hudson. Raven Press, New York © 1979.

Partitioning of Intramembrane Particles During the Freeze-Fracture Procedure

Birgit H. Satir and Peter Satir

Department of Anatomy, Albert Einstein College of Medicine, Bronx, New York 10461

This paper deals with the "partitioning" of intramembrane particles that takes place during the freeze-fracture procedure. As has been discussed in detail in this volume, when lipid bilayers are frozen to liquid nitrogen temperatures, they become thermodynamically unstable and tend to split along the middle of the hydrophobic bilayer, in such a manner that two complementary halves or two monolayers of lipids are obtained. As mentioned elsewhere, the two halves of a membrane produced upon fracturing, or two monolayers, are referred to as E, the exterior half, or the exterior monolayer, and the P, protoplasmic half, or protoplasmic monolayer, of the membrane. Each monolayer has sides consisting of one of the original membrane surfaces (the PS and ES, respectively) and one of the newly revealed fracture faces (the PF and EF).

In an earlier paper (Satir and Satir, 1974), we dealt with the problem as to which half of the membrane the particle would adhere to after fracturing. We suggested that for single classes of particles, defined by size and usually by equivalence of position in a charcteristic array, a particle partition coefficient (Kp) can be calculated such that

$$(1) \quad Kp = C_P/C_E,$$

where C_P and C_E are the concentrations in numbers of particles per unit surface adhering to the P and E faces, respectively. We demonstrated that in one array, the *Tetrahymena* fusion rosette, we could account for the variability in array appearances on a fracture face if we assume that the fracture can pass to either side of an intercalated particle via a stochastic process, dependent only on relative bond strength to either side of the intercalation. In this interpretation any given freeze-fracture particle can be found on either the P or the E fracture face, with a probability measured by the particle partition coefficient (Kp). Because important features of the particle can be located on either its protoplasmic or exterior side, Kp can be thought of as a measure of the interaction of the opposite sides of a particle with adjacent molecules, which of course can change with specimen preparation.

Since this concept was proposed, it has become clear that neither the particle nor the partition coefficient behaves quite as simply during fracturing as was first thought. The picture is still far from complete, but in this article we summarize some of the complications now known. As will become evident, these complications are not totally unexpected because they correspond to a more current, and we hope, a more realistic picture of the cell membrane.

THE COMPOSITION OF INTRAMEMBRANE PARTICLES

Originally intramembrane particles were proposed to correspond to integral proteins or lipoprotein complexes (see Branton and Deamer, 1972, for a summary of original evidence). It seemed apparent that the most simple explanation— one particle, one polypeptide—was not correct because the particle size 7–8 nm was too great to correspond to single polypeptides. Some of the most telling evidence for a protein-particle relationship was that of Pinto da Silva et al. (1970) which indicated that freeze-fracture particles in the erythrocyte membrane corresponded to or at least moved together with antibody receptor sites for IgA. However, these experiments are subject to certain interpretational difficulties and in addition, are perhaps not able to be entirely generalized. For example, Robertson proposed the alternative explanation that the particles were aggregations of specific lipids organized under appropriate peripheral proteins (Robertson, 1977). This would be consistent with the earlier finding of Pinto da Silva and Branton that ferritin, covalently bonded to membrane proteins at the cell surface, is not seen in the fracture face. The conventional view is that the fracture plane follows the path of least resistance—that is, covalent bonds are not broken. This interpretation was also used by Pinto da Silva and Branton (1970) in their study of the location of the fracture plane. Human erythrocyte ghosts were labeled with ferritin covalently bound to the membranes with the bifunctional reagent toluene-2,4 diisocyanate (TC) and subsequently fractured. When the TC-conjugating agent was used, ferritin was always present on both etched faces and not present on either fracture face. However, if covalent bonds were not broken during fracturing, one should expect to see the ferritin on the fracture faces and not only on the surfaces. Several workers have attempted to distinguish between these and other proposed models of intramembrane particles (IMP) by direct tests with reconstitution systems where the lipid bilayer composition and the added proteins could be strictly defined. The following three systems are noteworthy.

(a) Purple membrane: this system has been examined in considerable detail by negative stain, electron diffraction (Unwin and Henderson, 1975), and freeze-fracture (Blaurock and Stoeckenius, 1971). A molecular picture has been obtained that suggests that in the purple membrane a freeze-fracture particle is the functional transmembrane protein pathway. It consists of bacterial rhodopsin in the form of seven α-helices arranged around a central core (Henderson and Unwin, 1975). The particle is asymmetrically set in the membrane such that in the natural system, the pump operates in one direction only and the particles fracture almost entirely with the P face ($Kp \gg 1$). As elsewhere, when the system is reconstituted, the asymmetry is destroyed. The particles are found equally on both P and E fracture faces ($Kp \sim 1$) and the pump is bidirectional.

(b) Ca-ATPase of the sarcoplasmic reticulum (SR): this is almost exactly analogous to the purple membrane. The Ca^{2+} pump of the SR has been isolated and characterized by MacLennan and colleagues (1971, 1973). They showed that reinsertion of the isolated enzyme into liposomes whose fracture faces are entirely smooth causes reappearance of particles. In the muscle cell, the particles adhere to the P-fracture face with a large Kp; in the reconstituted system the Kp drops considerably, and again the pump is bidirectional.

(c) Human erythrocyte MN-glycoprotein with phospholipid bilayers: Segrest et al. (1974) have studied the kinetics of reappearance of 8 nm IMP in phospholipid bilayers incubated with a membrane penetrating hydropholic tryptic peptide (Tis) from the major surface glycoprotein of the human erythrocyte. The results show that above a critical multimer concentration these small polypeptides come together within the lipid bilayer to give rise to an IMP. Below this concentration no IMPs are seen.

From these studies, we can conclude that in many cases IMPs correspond to arrays of identifiable proteins, sometimes with obvious transmembrane functions. A rough correspondence between asymmetry of insertion and function versus particle partition coefficient is found, suggesting that the original interpretation of Kp is useful for a single identifiable class of IMPs.

DOES THE FRACTURE CLEAVE COVALENT BONDS?

Recently, Edwards et al. (1979) reported an interesting result by using erythrocytes and the freeze-fracture technique. They isolated half membranes of the erythrocytes by using the method of Fisher (1975). After fracturing, the membranes were solubilized in SDS and analyzed by SDS-PAGE. Because the SDS-PAGE profile of the erythrocytes is well known, they could look directly for the appearance of protein fragments caused by cleavage of covalent bonds in the membrane during the fracture process. They found four small polypeptide fragments in the E membrane halves. These fragments indicate that the transmembrane sialoglycoproteins were fragmented and that freeze-fracture can break covalent bonds. Because the fragments always appear as well defined molecular weights, the fracture has not taken place randomly but may occur at specific primary sequence sites. Band 3 remained largely with the P half of the membrane, consistent with the idea that band 3 is associated with proteins located at the cytoplasmic surface. In summary, freeze-fracture may affect integral proteins in two different ways: (a) intact proteins may be "pulled through" with the external half or pulled out upon fracturing, i.e., proteins are subject to plastic deformation at the moment of fracture; (b) covalent bonds may be broken. This depends on how the particular protein is "anchored" to either side of the membrane as described by the particle partition coefficient.

QUICK FREEZING VERSUS FIXATION

The concept of using particle partition coefficient (Kp) to describe intramembrane particle arrays grew out of a general paucity of useful definitions of particle arrays apart from the properties simply observed in freeze-fracture electron microscopy—that is, the shape, size, number, and configuration. We (Satir and Satir, 1974) previously discussed the significance of why, in many intramembrane particle arrays, Kp differed from unity: (a) single particles within an array may not be symmetrically located or specifically bonded to both sides of the membrane, also suggested by Speth and Wunderlich (1972); (b) during preparation for freeze-fracture, which normally involves cross-linking agents such as glutaraldehyde or various fixatives, changes might very well occur which may alter the bonding strength to either half of the membrane. As shown earlier by Dempsey et al. (1973) in capillary epithelium, after fixation, particles may fracture to the face opposite to their position in the "native" state; (c) different Kps represent different natural associations with either partner particles within an array or surface extensions to either half of the membrane.

Transmembrane proteins can now be shown to interact with either surface ligands such as multivalent concanavalin A (con A) at the ES or cytoskeletal elements (microtubules or microfilaments, probably indirectly) at the PS in such a way as might be predicted to alter Kp. For example, Condeelis (1979) has directly demonstrated an association between ES con A receptors and PS actin and myosin across a cell membrane during capping in *Dictyostelium*. How ephemeral this association is during fracturing of the membrane and where and whether IMPs are appropriately found has not yet been determined. Kp measured before versus after fixation in such situations may be useful in mapping membrane-cytoskeletal interactions. We might, for instance, expect Kp to change upward as the submembrane microfilament associations become tighter or more extensive.

Recently, it has become possible to approach the "native" state of the membrane by using what is known as "quick freezing" (Ornberg and Reese, *this volume*). An example of this procedure and its effect on the Kps for the fusion rosette in *Paramecium* is shown with (Fig. 1) and without (Fig. 2) fixation and cryoprotection (Satir and Heuser, unpublished).

In Fig. 1, as has been previously shown (Beisson et al., 1976), the majority of the fusion rosette particles are found associated with the PF of the membrane and hardly any with the EF. In the quick freezing method (Fig. 2) (here printed to show black shadow), the reverse appears to be the case. A similar result was also reported by Lefort-Tran et al. (1978). We compare the Kps of the *Paramecium* fusion rosette measured in several instances in Table 1.

Although the Kp of the fusion rosette particles clearly reverses with fixation, the Kp of the IMPs associated with the ring does not. We believe that reversal occurs because fixation alters specific associations of the rosette particles at the ES and PS; if this is the case, surprisingly, we anticipate that associations

PF   **EF**

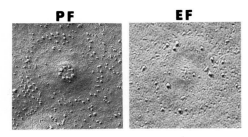

CONVENTIONAL METHOD
FIXED
CRYOPROTECTED

1

FIGS. 1 and 2. Illustrations of the secretory site in *Paramecium* as seen from the protoplasmic and exterior halves of plasma membrane with (Fig. 1, conventional method) or without (Fig. 2, quick freezing) fixation and cryoprotection. The site consists of the central fusion rosette surrounded by one or two outer rings of particles. Note the change in rosette particles from PF (with fixation) to EF (no fixation, quick freezing). The shadow is shown black in Fig. 2, white in Fig. 1.

PF   **EF**

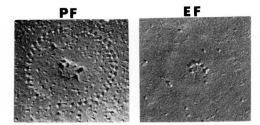

QUICK FREEZING
NO FIXATION
NO CRYOPROTECTANT

2

TABLE 1. *Particle partition coefficient of the Paramecium fusion rosette*

Cell	Growth temp. °C	Fixation	Total particles per rosette (Mean no.) PF + EF	Kp
wt	18	+	8	2.9 [b]
nd9	18	+	6	2.9 [b]
wt	27	+	10	5.0 [b]
wt	. . [a]	−	8	0.17

[a] Room temperature ($>18°<27°C$).
[b] Beisson et al. (1976).

(with the cytoskeleton?) at the PS are strengthened by fixation and/or those at the ES (charge interactions in the cell coat?) are weakened. The total number of particles (PF and EF) seen per rosette is, as anticipated, unaffected by fixation.

EFFECT OF ASSOCIATION STATE AND TEMPERATURE ON K_p

Although Kp was formulated as a constant, regardless of association state of the IMPs, the original data did not actually demonstrate this point. Ojakian and Satir (unpublished results; see Ojakian, 1974) showed that in the chloroplast thylakoid membrane when association state was varied for several particle classes, Kp was constant for only a single class (7-nm diameter particles). For three classes of particles (10.5, 14, and 16-nm diameters) Kp varied with association state, as might be expected if individual IMPs interact as they associate.

Table 1 indicates that in the *Paramecium* fusion rosette Kp is relatively independent of total particle number, under the same conditions of temperature and fixation. It is interesting to note, however, that Kp may change somewhat with temperature.

Clearly, these experiments are highly preliminary and Kp as a concept is subject to some important reservations in practical use. However, it now seems worthwhile to utilize variable conditions of fixation versus quick freezing, external ligands that cross-link the cell coat, agents that affect cytoskeletal assembly state, different association states of IMPs, temperature, etc., in a systematic study of particle partition coefficient in certain well defined arrays such as the fusion rosette as a means of relating the transmembrane protein assemblies that comprise IMPs to physiological events.

ACKNOWLEDGMENT

This work was supported by grants from U.S. Public Health Service GMS 24724 and HL 22560.

REFERENCES

1. Beisson, J., Lefort-Tran, M., Pouphile, M., Rossignol, M., and Satir, B. (1976): Genetic analysis of membrane differentiation in *Paramecium*. Freeze-fracture study of the trichocyst cycle in wild-type and mutant strains. *J. Cell Biol.,* 69:126–143.
2. Blaurock, A. E., and Stoeckenius, W. (1971): Structure of the purple membrane. *Nature (New Biol.),* 233:229–232.
3. Branton, D., and Deamer, D. W. (1972): Membrane structure. Protoplasmatologia Series. Springer-Verlag, New York.
4. Condeelis, J. (1979): Isolation of concanavalin A caps during various stages of formation and their association with actin and myosin. *J. Cell Biol.,* 80:751–758.
5. Dempsey, G. P., Bullivant, S., and Watkins, W. B. (1973): Endothelial cell membranes: polarity of particles as seen by freeze-fracturing. *Science,* 179:190–192.
6. Edwards, H. H., Mueller, T. J., and Morrison, M. (1979): Distribution of transmembrane polypeptides in freeze fracture. *Science,* 203:1343–1346.
7. Fisher, K. A. (1975): "Half" membrane enrichment: verification by electron microscopy. *Science,* 190:983–985.

8. Henderson, R., and Unwin, P. N. T. (1975): Three-dimensional model of purple membrane obtained by electron microscopy. *Nature*, 257:28–32.
9. Lefort-Tran, M., Gulik, T., Plattner, H., Beisson, J., and Wiessner, W. (1978): Influence of cryofixation procedures on organization and partition of intramembrane particles. Ninth International Cong. on Electron Microscopy, II:146.
10. MacLennan, D. H., Seeman, P., Iles, G. H., and Yip, C. C. (1971): Membrane formation by the adenosine triphosphatase of sarcoplasmic reticulum. *J. Biol. Chem.*, 246:2702–2710.
11. MacLennan, D. H., Yip, C. C., Iles, G. H., and Seeman, P. (1973): Isolation of sarcoplasmic reticulum proteins. *Cold Spring Harbor Symp. Quant. Biol.*, 37:469–477.
12. Ojakian, G. K. (1974): Structure and fluidity in *Chlamydomonas* chloroplast membranes. Ph.D. thesis, University of California, Berkeley.
13. Pinto da Silva, P., and Branton, D. (1970): Membrane splitting in freeze-etching. Covalently bound ferritin as a membrane marker. *J. Cell Biol.*, 45:598–605.
14. Pinto da Silva, P., Douglas, S. D., and Branton, D. (1970): Localization of A antigen sites on human erythrocyte ghosts. *Nature*, 232:194–196.
15. Robertson, J. D. (1977): On the nature of intramembrane particles. 35th Ann. Proc. Electron Microscopy Soc. America, pp. 680–683.
16. Satir, P., and Satir, B. (1974): Partition co-efficient of membrane particles in the fusion rosette. *Exp. Cell Res.*, 89:404–407.
17. Segrest, J. P., Gulik-Krzywicki, T., and Sardet, C. (1974): Association of the membrane-penetrating polypeptide segment of the human erythrocyte MN-glycoprotein with phospholipid bilayers. I. Formation of freeze-etch intramembranous particles. *Proc. Natl. Acad. Sci. USA*, 71:3294–3298.
18. Speth, V., and Wunderlich, F. (1972): Membranes in *Tetrahymena:* I. The cortical pattern. *J. Ultrastruc. Res.*, 41:258–269.
19. Unwin, P. N. T. and Henderson, R. (1975): Molecular structure determination by electron microscopy of unstained crystalline specimens. *J. Mol. Biol.*, 94:425–440.

Note added in proof: Verkleij et al. (*Nature,* 279:162–163) report that lipid can form IMPs under appropriate conditions.

Pretreatment Artifacts in Plant Cells

*J. H. Martin Willison and **R. Malcolm Brown, Jr.

* Biology Department, Dalhousie University, Halifax, Nova Scotia, B3H 4J1, Canada;
and ** Botany Department 010-A, University of North Carolina,
Chapel Hill, North Carolina 27514

Among the principal differences between plant and animal cells is the possession by the plant cell of a substantial cell wall. Under normal conditions, most plant cells are in a state of "turgor" in which the limiting plasma membrane of the cell is pressed tightly against the cell wall. We will concentrate upon artifacts induced at this plasma membrane–cell wall interface by pretreatments with glycerol or glutaraldehyde.

The materials and methods used in the present study have been described previously and are only summarized here. *Phaseolus vulgaris* (bean) root tips were excised and placed directly in 20% (v/v) glycerol–water, or were frozen in a small drop of water immediately after excision (14,15). The unicellular alga, *Oocystis apiculata* was frozen in its culture medium either with or without fixation in glutaraldehyde or glutaraldehyde/tannic acid (6). Enzymatically isolated tomato protoplasts (19) and tobacco mesophyll protoplasts (18) were cultured in defined media and freeze etched either in a culture medium containing 20% sucrose or after fixation in 6% glutaraldehyde in 0.05 M phosphate-buffered 20% sucrose at pH 7. Freeze-etch replicas were prepared using a Balzers 360 M freeze-etch apparatus and the conventional procedure described by Moor and Mühlethaler (7).

In most early studies of plant cell structure by freeze etching, tissues were infiltrated with glycerol solutions to circumvent the problem of modification of cellular ultrastructure as a result of ice segregation. Infiltration of glycerol was achieved by "growing" the material (usually roots) in glycerol solutions (see for examples: 1,4,9,12), commonly by passing the material through steadily increasing concentrations of the agent over periods of several days (4,12). As in studies of animal cells during this period, relatively little attention was paid to the possibility that glycerol might induce structural artifacts, although it was noted that growth virtually ceased (4), cytoplasmic streaming stopped (11), and evidence of plasmolysis could be found (4). Nevertheless, even after treatment with glycerol concentrations above 30% for several days, cytoplasmic streaming may be reinitiated in certain tissues once the glycerol has been removed (11).

In recent comparative studies of plasma membranes in untreated and glycerol-treated plant tissues (14,15), structurally deleterious effects of treatment with glycerol have been noted. Within 1 min of treatment with 20% glycerol, a high degree of plasmolysis occurs (Fig. 1). After exposure for 30 min or more, penetration of glycerol into the cells gives rise to deplasmolysis, and freeze-etch micrographs of the type widely reported for glycerol-treated material may be obtained (15). In particular, the plasma membrane is always relatively smooth and the openings to plasmodesmata are funnel shaped (Fig. 2). Distinctive structural modifications, having the appearance of "flow lines" or "rivulets" (Fig. 3), may be found occasionally in plasma membranes after the tissue has been treated with glycerol (15). By contrast, in material that has been freeze fractured after having been frozen without pretreatment, the plasma membrane appears to be pressed so tightly against the cell wall that the structure of the inner surface of the cell wall is clearly visible as an impression in the plasma membrane (Fig. 4), and "flow lines" are absent. Furthermore, the sites of plasmodesmata appear as slight mounds having membrane particles clustered in the center (Fig. 4) (16). The condition shown in Fig. 4 is to be expected in normal plant cells that are under turgor pressure (14). The smooth fracture faces of plasma membranes in glycerol-treated material indicate that turgor is not regained, and, because turgor is required for growth, the lack of root growth in glycerol solutions (4) is probably due to the lack of turgor resulting from glycerol-induced plasmolysis. Other changes induced by glycerol treatment of plant tissues have not been described, although it is probable that intracellular membrane vesiculation, of the type described in animal cells by Moor (8) occurs after long periods of incubation in glycerol solutions (see micrographs in refs. 4 and 13).

Prefixation with glutaraldehyde offers a solution to the problems of structural modifications resulting both from ice segregation and glycerol treatment, and it is clearly a useful control treatment in general cytological studies. However, in using the freeze-fracture technique it is important to realize that glutaraldehyde fixation may induce significant changes, particularly in the fracturing of biological membranes. This problem is exemplified in the studies described below.

Plant protoplasts, isolated by the enzymatic digestion of the cell walls from a variety of plant tissues, are currently important experimental systems in plant biology (3). Isolated protoplasts may be cultured in hypertonic media and regenerate a cell wall after a period of time. The effects of glutaraldehyde on the "particle partition coefficients" of fractured plasma membranes of isolated plant protoplasts are illustrated in Figs. 5 to 7 (13). When protoplasts are freshly isolated and frozen in the culture medium without prefixation, the plasma membranes fracture in such a way that the particle density on the E fracture face slightly exceeds that on the P fracture face (Figs. 5 and 6) (13,18). When similar isolated protoplasts (either from tobacco mesophyll or from tomato locule tissue) are glutaraldehyde fixed, a marked inequality of particle partition is found, with the P fracture face of the plasma membrane having a much higher particle density than the E fracture face (Figs. 7 and 8), both when glutaraldehyde fixation is followed by glycerol treatment or when the fixed protoplasts are

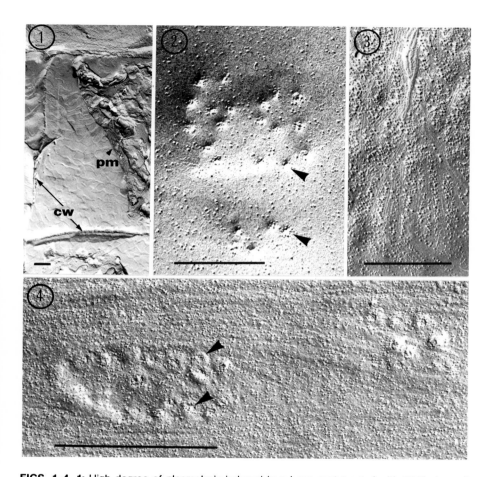

FIGS. 1–4. 1: High degree of plasmolysis induced in a bean root treated with 20% glycerol for 1 min; cell wall (cw); plasma membrane (pm). Bar 1 μm. × 4,300. **2:** E fracture face of the plasma membrane of a cell in the tip of a bean root treated with 20% glycerol for 20 min. Note that the membrane is relatively smooth and that the openings to plasmodesmata *(arrows)* are funnel shaped. Bar 500 nm. × 48,000. **3:** P fracture face of the plasma membrane of a cell from the tip of a bean root treated with 20% glycerol for 20 min. The "flow lines" are typical of those that occur occasionally in glycerol-treated material. Bar 500 nm. × 47,000. **4:** E fracture face of the plasma membrane of a cell from the tip of a bean root frozen without pretreatment. Notice that impressions made by the microfibrils of the underlying wall are clear and that the openings to plasmodesmata *(arrows)* are surrounded by a raised annulus as might be expected. Bar 1 μm. × 43,000.

frozen in the absence of glycerol. Interestingly, a similar (though lesser) inequality of partition, that is, particle density on P-face (PF) exceeds particle density on E-face (EF), is found in the plasma membranes of normal plant cells in tissues (Figs. 9 and 10) and in isolated plant protoplasts, once cell wall regenera-

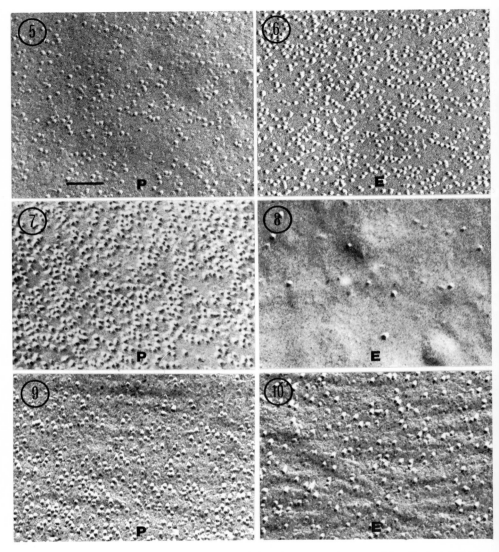

FIGS. 5–10. 5 and 6: P and E fracture faces of the plasma membranes of freshly isolated plant protoplasts frozen in their culture media. Bar (Fig. 5) 100 nm. **7 and 8:** P and E fracture faces of the plasma membranes of freshly isolated plant protoplasts fixed in glutaraldehyde before freezing. **9 and 10:** P and E fracture faces of the plasma membranes of tobacco leaf cells from a tissue piece frozen without pretreatment. Figs. 5–10: × 98,000.

tion has been initiated (18). These results demonstrate that changes may be found in particle-partition coefficients of plasma membranes, that correlate with biological changes in plant cells. However, since glutaraldehyde fixation itself induces changes in membrane particle-partition coefficients, the meaningfulness of any such partition coefficients in glutaraldehyde-fixed material remains doubtful.

The effect of glutaraldehyde on particle partition between plasma-membrane fracture moieties is not restricted to isolated plant protoplasts, as demonstrated by the following observations on the unicellular green alga *Oocystis apiculata*. Freeze-fractured plasma membranes of cells of *Oocystis* that are actively depositing a cell wall contain ordered complexes of membrane-associated particles which are considered to be important both in synthesizing cellulosic microfibrils and in assuring the development of an orderly cell wall (2,6). When no pretreatment has been used, the P fracture face (Fig. 11) bears "granule bands" where the newly synthesized microfibrils of the cell wall lie tightly pressed to the plasma membrane, and the E fracture face (Fig. 12) bears "terminal complexes" that are associated with the ends of microfibrils where microfibril synthesis is consid-

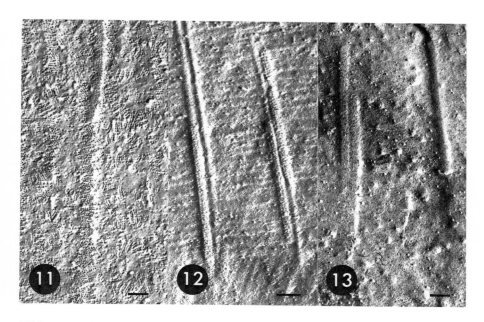

FIGS. 11–13. P fracture face (Fig. 11) and E fracture faces (Figs. 12 and 13) of plasma membranes of *Oocystis apiculata* during the phase of cell-wall deposition. The particle segregation between P face "granule bands" (Fig. 11 × 57,000) and E face "terminal complexes" (Fig. 12, × 53,900) is typical of untreated material. After fixation with glutaraldehyde–tannic acid (Fig. 13, × 57,000), although the loci of "terminal complexes" are visible, the granular complexes themselves are absent. Bar 0.1 μm. (Fig. 13 from Montezinos and Brown, ref. 6, with permission of Alan R. Liss, Inc.)

ered to occur. After glutaraldehyde pretreatment (both with or without tannic acid in the fixative solution) the segregation between the two fracture faces is altered leaving the E fracture face devoid of "terminal complexes" (Fig. 13).

The causes of these changes in particle-partition coefficients can only be speculated upon at present. Glutaraldehyde may act to increase the bonding of intracellular components (notably proteins) to membrane proteins, thereby increasing the tendency for membrane proteins to be retained by the protoplasmic (P) rather than the exoplasmic (E) half of the membrane after fracturing. When isolated protoplasts are generated by removing extracellular polysaccharide components that might be expected to be bonded (at least partially) to membrane intercalated proteins, the tendency for membrane particles to be retained by the exoplasmic half of the membrane appears to be increased.

The results described above demonstrate clearly that pretreating plant cells with either glycerol or glutaraldehyde can have deleterious effects upon plasma membrane structure and upon plasma-membrane particle partition. Plasma membranes often appear to be well preserved in material that is frozen without pretreatment, despite the presence of intracellular ice crystals. Nevertheless, ice crystals do arise sometimes at the cell wall–plasma membrane interface in plant tissues (15) and, as in studies of intracellular organization, rapid freezing rates are required if these are to be eliminated. Very fast freezing, as for example in spray freezing (10), is inevitably limited to isolated cells, cell monolayers, or to the peripheral cells of tissue pieces, because of the relatively poor thermal conductivity of aqueous materials. The idea of freezing the specimen without pretreatment should not be discarded out-of-hand, however. Despite the presence of intracellular ice, valuable results concerning intracellular organization have been obtained from untreated plant materials (5,17).

On the basis of the discussion above, the following conclusions may be drawn:

(a) The deleterious effects of short-term glycerol treatment center around the induced plasmolysis; observations on plasma-membrane topology and substructure in unfixed glycerol-treated material is of doubtful value.

(b) Glutaraldehyde fixation causes major changes in the particle-partition coefficients of plasma membranes of both isolated plant protoplasts and the alga *Oocystis* (in the absence of evidence to the contrary, it should be assumed that this is a general phenomenon). Adequate unfixed controls should therefore be used in studies of plasma membrane substructure.

(c) Despite the presence of intracellular ice crystals, plasma-membrane structure and the relationship between the plasma membrane and the cell wall appear to be well preserved in many plant tissues that have been frozen without pretreatment.

ACKNOWLEDGMENTS

The electron micrographs shown in this work were prepared in three laboratories: Botany Department, University of Nottingham; Microbiology Department,

Dalhousie University; and Botany Department, University of North Carolina. We are grateful to Prof. E. C. Cocking (Nottingham) and Dr. K. B. Easterbrook (Dalhousie) for the provision of these facilities. The micrographs of *Oocystis* were prepared by Dr. David Montezinos, to whom we are particularly grateful. This work was supported by grants from the National Research Council of Canada (to J. H. M. Willison) and the National Science Foundation (to R. M. Brown).

REFERENCES

1. Branton, D., and Moor, H. (1964): Fine structure in freeze-etched *Allium cepa* L. root tips. *J. Ultrastruct. Res.,* 11:401–411.
2. Brown, R. M., Jr., and Montezinos, D. (1976): Cellulose microfibrils: visualization of the biosynthetic and orienting complexes in the plasma membrane. *Proc. Natl. Acad. Sci. USA,* 73:143–147.
3. Cocking, E. C. (1972): Plant cell protoplasts, isolation and development. *Ann. Rev. Plant Physiol.,* 23:29–50.
4. Fineran, B. A. (1970): The effects of various pre-treatments on the freeze-etching of root tips. *J. Microsc.,* 92:85–97.
5. Johnson, R. P. C. (1973): Filaments but no membranous transcellular strands in sieve pores in freeze-etched, translocating phloem. *Nature,* 244:464–466.
6. Montezinos, D., and Brown, R. M., Jr. (1976): Surface architecture of the plant cell: Biogenesis of the cell wall, with special emphasis on the role of the plasma membrane in cellulose biosynthesis. *J. Supramol. Struct.,* 5:277–290.
7. Moor, H., and Mühlethaler, K. (1963): Fine structure of frozen etched yeast cells. *J. Cell Biol.,* 17:609–628.
8. Moor, H. (1971): Recent progress in the freeze-etching technique. *Philos. Trans. R. Soc. Lond. (Ser. B),* 261:121–131.
9. Northcote, D. H., and Lewis, D. R. (1968): Freeze-etched surfaces of membranes and organelles in the cells of pea root tips. *J. Cell Sci.,* 3:199–206.
10. Plattner, H., Fischer, W. M., Schmitt, W. W., and Bachmann, L. (1972): Freeze etching of cells without cryoprotectants. *J. Cell Biol.* 53:116–126.
11. Richter, H. (1968): Die Reaktion hochpermeabler Pflanzenzellen auf drei Gefrierschutzstoffe (Glyzerin, Äthylenglykol, Dimethylsulfoxid). *Protoplasma,* 65:155–166.
12. Rottenburg, W., and Richter, H. (1969): Automatische Glyzerinbehandlung pflanzlicher Dauergewebszellen für die Gefrierätzung. *Mikroskopie* 25:313–319.
13. Willison, J. H. M. (1973): An investigation of the application of freeze-etching. Ph. D. Thesis, University of Nottingham, England.
14. Willison, J. H. M. (1975): Plant cell-wall microfibril disposition revealed by freeze-fractured plasmalemma not treated with glycerol. *Planta,* 126:93–96.
15. Willison, J. H. M. (1976a): An examination of the relationship between freeze-fractured plasmalemma and cell-wall microfibrils. *Protoplasma,* 88:187–200.
16. Willison, J. H. M. (1976b): Plasmodesmata: A freeze-fracture view. *Can. J. Bot.,* 54:2842–2847.
17. Willison, J. H. M. (1976c): The hexagonal lattice spacing of intracellular crystalline tobacco mosaic virus. *J. Ultrastruct. Res.,* 54:176–182.
18. Willison, J. H. M., and Cocking, E. C. (1975): Microfibril synthesis at the surfaces of isolated tobacco mesophyll protoplasts, a freeze-etch study. *Protoplasma,* 84:147–159.
19. Willison, J. H. M., Grout, B. W. W., and Cocking, E. C. (1971): A mechanism for the pinocytosis of latex spheres by tomato fruit protoplasts. *Bioenergetics* 2:371–382.

Freeze Fracture: Methods, Artifacts, and Interpretations, edited by J. E. Rash and C. S. Hudson. Raven Press, New York © 1979.

Extrusion of Particle-Free Membrane Blisters During Glutaraldehyde Fixation

Elizabeth D. Hay and *David L. Hasty

Department of Anatomy, Harvard Medical School, Boston, Massachusetts 02115

Particle-free membrane vesicles or so-called blisters (18) observed in freeze-fracture replicas have been implicated in a number of physiological processes in several recent investigations. It has been suggested, for example, that particle-free blisters on mast-cell surfaces are a stage in the excretion of intracellular granules (6,16). Particle-free "mounds" filled with vesicles have been implicated as a source of new membrane for nerve-growth cones (4,23). Thus, it was with some interest that we noted the occurrence of similar-appearing blisters lacking intramembranous particles (IMP) on the surfaces of corneal fibroblasts and in the extracellular matrix (ECM) surrounding the cells (7,8).

The particle-free blisters observed in freeze-fracture replicas of the developing avian cornea are of several types: (a) Particle-free blisters attached to the cell membrane are particularly common (Figs. 1 and 2). They are generally round or oblong in shape, varying from hundred to several hundred nanometers in diameter. Their bounding membrane is smooth with only a hint of an occasional particle, whereas the adjacent plasmalemma with which they are continuous is rich in IMP that increase in number with age. (b) A second type of IMP-free blister is a membrane-bounded vesicle seemingly floating free in the extracellular space (7,8). We believe these free blisters derive from the attached blisters because they are also poor in IMP and because transitions between the two kinds of vesicles can be observed. (c) A third type of particle-free membranous blister observed in these freeze-fracture replicas has the structure of a typical myelin figure, being composed of multiple lipid bilayers (8). (d) A fourth type of particle-free structure that can be seen in freeze-fracture replicas has the appearance of a mound containing multiple vesicles. The IMP-free surface membrane of these large blisters (diameter up to several thousand nanometers) may follow the contour of the underlying vesicles (Fig. 3) or the vesicles may be enclosed by the blister (8). The component vesicles (100 to 200 nm in diameter) are also IMP poor. Occasionally, the large blisters may be surrounded by pits

---

* Present address: Department of Anatomy, University of Tennessee Center for the Health Sciences, Memphis, Tennessee 38163.

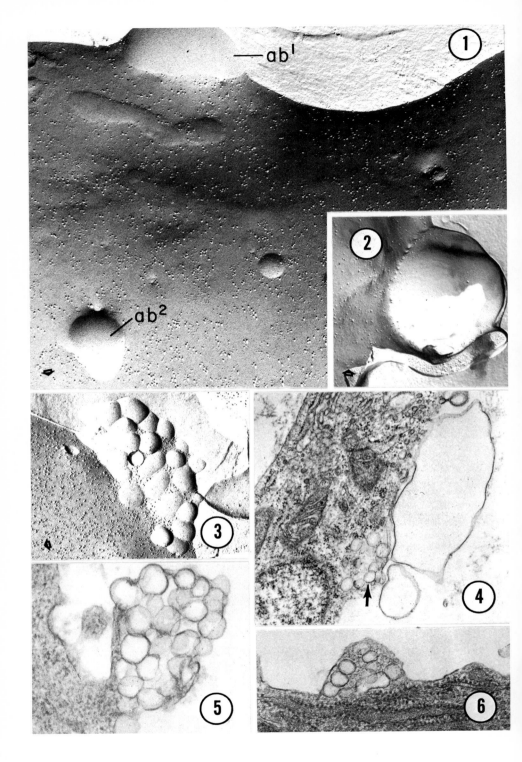

(Fig. 2; see also ref. 8) similar to those surrounding the mounds observed in nerve growth cones (see K. H. Pfenninger, *this volume*).

We examined thin sections in the electron microscope to look for corresponding structures in corneas that had been routinely fixed (2.5% glutaraldehyde and 1% formaldehyde for 30 min, followed by 1.5% OsO_4 in 0.1 M sodium cacodylate buffer, pH 7.4, with or without 0.01 M $CaCl_2$, for 30 min). The material we fractured was fixed in the same aldehyde formula (13), but was not postfixed in OsO_4, since the latter interferes with membrane cleavage (19); most of the material for freeze fracture was glycerinated, but the blisters appear whether or not glycerin is used.

In sections of glutaraldehyde-fixed corneas, structures corresponding to the attached and free blisters observed in fracture replicas are common. The bounding membrane is trilaminar in structure (8) and the blisters seem totally empty in content, even when attached to the plasmalemma (Fig. 4). Structures occur with morphology intermediate between the simple blisters and the multivesiculate blisters; this suggests that they may all have a similar origin from the cell membrane. Transitions between simple blisters and myelin figures can also be observed in sections. The myelin figures are composed of whorls of adherent trilaminar membranes (8). The tendency of blister membranes to adhere in this manner may be due to the fact that they are virtually IMP free, with resultant reduced repulsive charges.

The structure in sections that corresponds to the mounds observed by freeze fracture is a large bag (diameter up to several thousand nanometers) filled with vesicles (100 to 200 nm in diameter) and usually attached to the cell membrane (Figs. 5 and 6). The bounding membrane is trilaminar and continuous with the plasmalemma. The enclosed vesicles are empty in appearance (Figs. 5 and 6). Usually, there is little or no cytoplasm around the vesicles (Fig. 5).

FIGS. 1–6. 1: Electron micrograph of a replica of a freeze fracture through a corneal fibroblast in a 14-day-old embryo fixed in glutaraldehyde before freezing. The attached blister (ab[1]) at the top contains a few minute intramembranous particles, whereas the one (ab[2]) in the lower left cornea is virtually IMP free. The small arrow indicates the direction of the shadowing. ×55,000. **2:** Electron micrograph of a replica of a 14-day-old corneal fibroblast fixed in glutaraldehyde and freeze fractured. Pits *(arrow)* surrounding the base of this elevation are quite similar to pits reported around the bases of "mounds" in nerve-growth cones. ×46,000. **3:** Freeze-fracture replica of aldehyde-fixed 19-day-old cornea showing what is probably a stage in the development of a multivesicular blister. ×50,000. **4:** An electron micrograph of a thin section through the corneal stroma of a 14-day-old embryo fixed in glutaraldehyde followed by osmium. The membrane blisters seen in thin sections after aldehyde fixation correspond to those visualized in freeze fractures. In this micrograph, the area labeled by the arrow may be a developing multivesicular blister. A large, empty-appearing attached blister and a few seemingly unattached blisters are also present. ×23,000. **5:** An electron micrograph of a thin section showing a fibroblast process in a 14-day-old cornea fixed in aldehyde followed by osmium. The membrane of the multivesicular blister shown here is continuous with the plasmalemma of the fibroblast, but is interrupted by the membrane of the component vesicles in the blister itself. This blister is probably the counterpart of the one shown in Fig. 3. ×55,000. **6:** Electron micrograph of a small multivesicular blister on a corneal fibroblast fixed in aldehyde followed by osmium. ×41,000.

Empty-appearing blisters, myelin figures, and mounds of this type have been observed in thin sections ever since the introduction of glutaraldehyde as a prefix and they have long been suspected to be artifacts (1,10,20,27). To explore the possibility that these structures arise artifactually during aldehyde fixation of the embryonic cornea, we fractured frozen corneas that had not been fixed in aldehyde and we also examined sections of material that had been fixed simultaneously in a mixture of aldehyde and OsO_4 (10,32). The aldehyde–osmium mixture consists of a 1:1 solution of 5% glutaraldehyde and 1% OsO_4 in 0.2 M cacodylate buffer, mixed together immediately before use, and changed several times during a 30-min period. Frequent change to fresh mixtures is critical to prevent tissue explosion, even when ferrocyanide (14) is added to the fixative.

The IMP-poor blisters are virtually absent in unfixed fractured material (Fig. 7), and the corresponding, empty-appearing blisters viewed in sections are eliminated by OsO_4–aldehyde fixation (Fig. 8). Therefore, we conclude that in the cornea the IMP-free blisters viewed in freeze fracture, that seem empty in content in corresponding sections, are artifacts of fixation. Similar blisters observed by scanning electron microscopy (SEM) have been shown to be fixation artifacts (29). It is important in SEM as well as in freeze-fracture studies to distinguish such blisters from real "blebs," and therefore we suggest that the term bleb be used only to refer to rounded cytoplasmic processes containing cytoplasm and IMP (2,8).

We argue that the IMP-free blisters develop during aldehyde fixation in the following manner: Because lipid molecules in the membrane are probably in continuous flow (3), those not bridged to protein by aldehyde (17,18) could be expected to move away from the plasmalemma when membrane proteins are immobilized by aldehyde fixation. The released lipids form vesicles because they cannot diffuse freely in an aqueous environment. In support of this theory, it should be noted that glutaraldehyde seems to cross-link membrane proteins via lysine (11), without reducing the fluidity of the lipid bilayers (12). When isolated cells are fixed by glutaraldehyde alone, many lipids are lost in subsequent washes (15,25) probably in the form of free blisters. Osmium tetroxide reduces the loss of neutral lipids (15). When added to cells before or simultaneously with glutaraldehyde it would be expected to fix the relative positions of amphipathic proteins and to immobilize molecular motion in the lipid bilayer (12). The action of the combined fix, thus, probably takes advantage of the superior immobilization of lipid by OsO_4 and of proteins by glutaraldehyde.

The formation of particle-free blisters during aldehyde fixation in the absence

FIGS. 7 and 8. 7: Electron micrograph of a freeze-fracture replica of an unfixed 9-day-old fibroblast. The surface is uneven and some IMP clumping occurs, but no blisters are observed. ×55,000. **8:** An electron micrograph of a thin section demonstrating the effect of fixation of the 14-day-old cornea with a combined glutaraldehyde–osmium tetroxide mixture (CP, cell process). There are no membranous blisters in preparations treated with osmium before or with aldehyde. ×27,000. *Inset:* The cytoplasmic vesicles (V) that are associated with the plasmalemma in these preparations have a moderately dense content and are probably secretory in nature. ×68,000.

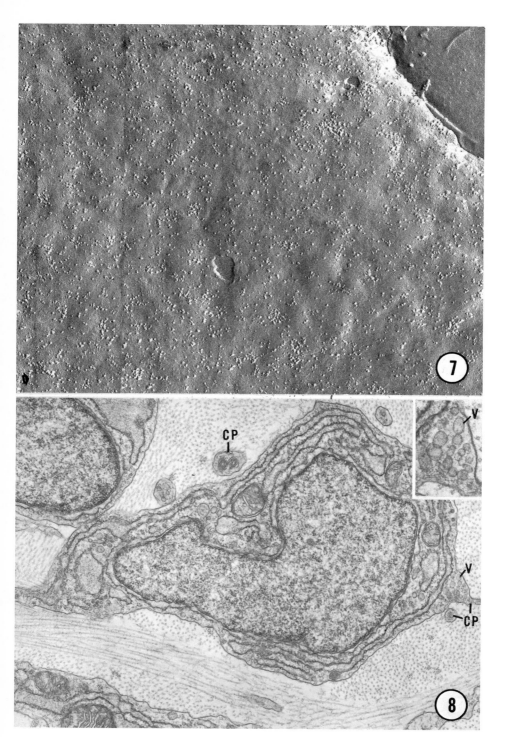

of OsO_4 draws dramatic attention to the fluidity (3) of the lipid layer. Long after the corneal fibroblasts are "dead" in the sense that their proteins are aldehyde denatured, some lipids continue to move as if "alive." A similar situation seems to hold in the case of certain phospholipid vesicles that are able to fuse with aldehyde-fixed cells and, here again, OsO_4 reduces membrane fluidity in the sense that it interferes with membrane fusion (24). Chylomicron triacylglycerol can be hydrolyzed by lipoprotein lipase in aldehyde-fixed specimens, the products seeming to take on the appearance of myelin figures (28). Labeled lipid can also move from one aldehyde-fixed cell to another (25). Aldehyde fixation, moreover, does not impair the ability of Sendai virus to fuse seemingly "dead" cells (31).

The postfixation membrane mobility observed in developing corneal fibroblasts has no obvious physiological significance. Blister-formation persists in mature corneas and thus probably does not reflect lipid instability owing to active ECM secretion, new membrane formation, or the relative IMP paucity of young membranes (7,8). It is conceivable, however, that postfixation lipid mobility has physiological significance in certain systems. Schneeberger et al. (26) observed blisters and myelin figures in aldehyde-fixed mitochondria of fibroblasts fed an excess of polyunsaturated acids. The blisters were prevented from forming by OsO_4 and thus are artifacts, but since they were not observed in normal mitochondria, their formation may reflect a meaningful membrane change (26). Stolinski et al. (30) in a recent freeze-fracture study reported that unfixed glycerinated chick embryo cells lack surface blisters, whereas glutaraldehyde added after glycerination induces blisters. They speculated that aldehyde fixation reveals sites of membrane accretion, but they offered no evidence for the idea.

Because the tendency to form mounds characterizes certain parts of the nerve-growth cone (4,5,22,23) they could conceivably represent areas of membrane accretion or instability, even though they probably also are fixation artifacts, at least in part (19a; but see also K. H. Pfenninger *this volume*). The same may prove to be true for mast-cell blisters (16). It will be important in future interpretations of particle-free blisters to compare the appearance of fractures and sections of rapidly frozen tissues (9) with fractures and/or sections of unfixed tissues, tissues prepared for microscopy by routine glutaraldehyde "prefixation," and tissues fixed with the combination of OsO_4 and glutaraldehyde we used in the present study.

ACKNOWLEDGMENT

Supported by grant HD-00143 from the United States Health Service.

REFERENCES

1. Arborgh, B., Bell, P., Brunk, U., and Collins, V. P. (1976): The osmotic effect of glutaraldehyde during fixation. A transmission electron microscopy, scanning electron microscopy and cytochemical study. *J. Ultrastruc. Res.,* 56:339–350.

2. Bard, J. B. L., Hay, E. D., and Meller, S. M. (1975): Formation of the endothelium of the avian cornea: A study of cell movement *in vivo. Develop. Biol.,* 42:334–361.
3. Bretscher, M. S. (1976): Directed lipid flow in cell membranes. *Nature,* 260:21–23.
4. Bunge, M. B. (1973): Fine structure of nerve fibers and growth cones of isolated sympathetic neurons in culture. *J. Cell Biol.,* 56:713–735.
5. Bunge, M. B. (1977): Initial endocytosis or peroxidase of ferritin for growth cones of cultured cells. *J. Neurocytol.,* 6:407–439.
6. Chi, E. Y., Lagunoff, D., and Koehler, J. K. (1976): Freeze-fracture study of mast cell secretion. *Proc. Natl. Acad. Sci. USA,* 73:2823–2827.
7. Hasty, D. L., and Hay, E. D. (1977): Freeze-fracture studies of the developing cell surface. I. The plasmalemma of the corneal fibroblast. *J. Cell Biol.,* 72:667–686.
8. Hasty, D. L., and Hay, E. D. (1978): Freeze-fracture studies of the developing cell surface. II. Particle-free membrane blisters on glutaraldehyde-fixed corneal fibroblasts are artefacts. *J. Cell Biol.,* 78:756–768.
9. Heuser, J. E., Reese, T. S., and Landis, D. M. D. (1975): Preservation of synaptic structure by rapid freezing. *Cold Spring Harbor Symp. Quant. Biol.,* 40:17–24.
10. Hirsch, J. G., and Fedorko, M. E. (1968): Ultrastructure of human leucocytes after simultaneous fixation with glutaraldehyde and osmium tetroxide and "postfixation" in uranyl acetate. *J. Cell Biol.,* 38:615–627.
11. Hopwood, D. (1972): Theoretical and practical aspects of glutaraldehyde fixation. *Histochem. J.,* 4:267–303.
12. Jost, P., Brooks, U. J., and Griffith, O. H. (1973): Fluidity of phospholipid bilayers and membranes after exposure to osmium tetroxide and glutaraldehyde. *J. Mol. Biol.,* 76:313–318.
13. Karnovsky, M. J. (1965): A formaldehyde–glutaraldehyde fixative of high osmolality for use in electron microscopy. *J. Cell Biol.,* 27:137a.
14. Karnovsky, M. J. (1971): Use of ferrocyanide-reduced osmium tetroxide in electron microscopy. In: *Abstracts of the 11th Annual Meeting of the American Society for Cell Biology,* New Orleans, La. p. 146. Rockefeller University Press, New York.
15. Korn, E. D., and Weisman, R. A. (1966): Loss of lipids during preparation of amoebae for electron microscopy. Biochim. *Biophys. Acta,* 116:309–316.
16. Lawson, D., Raff, M. C., Gomperts, B., Fewtrell, C., and Gilula, N. B. (1977): Molecular events during membrane fusion. A study of exocytosis in rat peritoneal mast cells. *J. Cell Biol.,* 72:242–259.
17. Litman, R. B., and Barrnett, R. J. (1972): The mechanism of the fixation of tissue components by osmium tetroxide via hydrogen bonding. *J. Ultrastruc. Res.,* 38:63–86.
18. McIntyre, J. A., Gilula, N. B., and Karnovsky, M. J. (1974): Cryoprotectant-induced redistribution of intramembranous particles in mouse lymphocytes. *J. Cell Biol.,* 60:192–202.
19. Nermut, M. V., and Ward, B. J., (1974): Effect of fixatives on fracture plane in red blood cells. *J. Microscopy,* 102:29–39.
19a. Nuttall, R. P., and Wessells, N. K. (1979): Veils, mounds, and vesicle aggregates in neurons elongating *in vitro. Exp. Cell Res. (in press).*
20. Olah, J., and Rohlich, P. (1966): Peculiar membrane configurations after fixation in glutaraldehyde. *Acta Biol. Acad. Sci. Hung.,* 17:65–73.
21. Pagano, R. E., and Huang, L. (1975): Interaction of phospholipid vesicles with cultured mammalian cells. I. Characteristics of uptake. *J. Cell Biol.,* 67:38–48.
22. Pfenninger, K. H. 1979. This volume.
23. Pfenninger, K. H., and Bunge, R. P. (1974): Freeze-fracturing of nerve growth cones in young fibers. A study of developing plasma membrane. *J. Cell Biol.,* 63:180–196.
24. Poste, G., and Papahadjopoulos, D. (1976): Lipid vesicles as carriers for introducing materials into cultured cells: Influence of vesicle lipid composition on mechanism(s) of vesicle incorporation into cells. *Proc. Natl. Acad. Sci. USA,* 73:1603–1607.
25. Poste, G., Porter, C. W., and Papahadjopoulos, D. (1978): Identification of a potential artefact in the use of electron microscope autoradiography to localize saturated phospholipids in cells. *Biochem. Biophys. Acta,* 510:256–263.
26. Schneeberger, E. E., Lynch, R. D., and Geyer, R. P. (1976): Glutaraldehyde fixation used to demonstrate altered properties of outer mitochondrial membranes in polyunsaturated fatty acid (PUFA) supplemented cells. *Exp. Cell Res.,* 100:117–128.
27. Scott, R. E. (1976): Plasma membrane vesiculation: A new technique for isolation of plasma membranes. *Science,* 194:743–745.

28. Scow, R. O., Blanchette-Mackie, E. J., and Smith, L. C. (1977): Role of lipoprotein lipase and capillary endothelium in the clearance of chylomicrons from blood: A model for lipid transport by lateral fusion in cell membranes. In: *Cholesterol Metabolism and Lipolytic Enzymes,* edited by J. Polonovski, pp. 143–164. Masson Publishing, New York.

29. Shelton, E., and Mowczko, W. E. (1978): Membrane blisters: A fixation artefact. A study in fixation for scanning electron microscopy. *Scanning* 1:166–173.

30. Stolinski, C., Breathnach, A. S., and Bellairs, R. (1978): Effect of fixation on cell membrane of early embryonic material as observed on freeze-fracture replicas. *J. Microscopy,* 112:293–299.

31. Toister, Z., and Loyter, A. (1973): The mechanism of cell fusion. II. Formation of chicken erythrocyte polykarons. *J. Biol. Chem.,* 248:422–432.

32. Trump, B. F., and Bulger, R. E. (1966): New ultrastructural characteristics of cells fixed in glutaraldehyde–osmium tetroxide mixture. *Lab. Invest.,* 15:368–379.

Freeze Fracture: Methods, Artifacts, and Interpretations, edited by J. E. Rash and C. S. Hudson. Raven Press, New York © 1979.

Scanning Electron Microscopy of Membrane Blisters Produced by Glutaraldehyde Fixation and Stabilized by Postfixation in Osmium Tetroxide

Emma Shelton and W. E. Mowczko

Department of Health, Education, and Welfare, Public Health Service, National Institutes of Health, Bethesda, Maryland 20205

Glutaraldehyde fixation produces membrane blisters on cells. Support for this statement is derived from light microscope observations of the fixation process and scanning electron micrographs of the fixed cells. Mouse peritoneal macrophages and lymphocytes, allowed to settle on glass cover slips in a perfusion chamber can be observed while being fixed in glutaraldehyde. Within 2 min after exposure to the fixative, many blisters form on the surface of the cells and an occasional one breaks off and floats away. Postfixation in OsO_4 produces subtle changes in the refractive index of the cell membrane but additional blisters cannot be identified positively. These experiments together with reports by Scott (2) and Hasty and Hay (1) led us to reconsider what we first believed to be a postfixation artifact produced by OsO_4 on cells prefixed in glutaraldehyde (3).

Cells that have been fixed in glutaraldehyde and postfixed in OsO_4, then further processed and examined in the scanning electron microscope have blisters on their surfaces (Fig. 1) almost without exception. Cells that have been fixed in glutaraldehyde or OsO_4 *alone* do not display these blisters (Fig. 2). Thus, the blisters appear to be an artifact of postfixation. In a lengthy examination of this phenomenon reported elsewhere (4), it could be shown that the formation of blisters is not a function of time, temperature, pH, or osmotic conditions during fixation and that it could not be prevented by the addition of tannic acid, chromium, or trinitroresorcinol to the fixative. It occurs on cells fixed *in vivo* as well as on cells adapted to long-term tissue culture. However, simultaneous fixation in glutaraldehyde and osmium tetroxide completely prevents blister formation, which rules out the assumption that the two fixatives act in some way to produce them. It seems most probable that after glutaraldehyde fixation the blisters are too delicate to survive the rigors of dehydration and critical-point drying and are broken off and lost, much as was observed in the light microscope. Thus, cells fixed solely in glutaraldehyde are free of blisters. Osmium

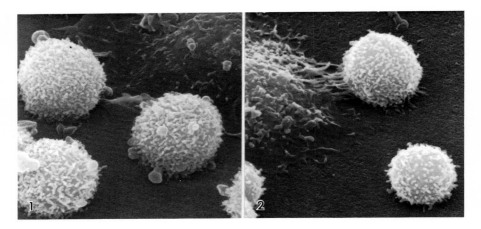

FIGS. 1 and 2. 1: Mouse peritoneal cells settled on Millipore filter. Numerous blisters decorate the surface of the lymphocytes and flattened macrophage. Cells were fixed in 5% glutaraldehyde for 60 min and postfixed in 1% OsO_4 for 30 min at room temperature. **2:** Mouse peritoneal cells allowed to settle on a Millipore filter. A flattened macrophage and two lymphocytes are illustrated. Cells were fixed in 5% glutaraldehyde for 60 min at room temperature. Figs. 1 and 2: ×3,600.

tetroxide fixation appears to stabilize these membrane structures so that postfixed cells are decorated with them.

When examined by transmission-electron microscope, blisters are shown to be formed by extensions of the cell membrane and, except for occasional whorls of membrane, to be empty (Fig. 3).

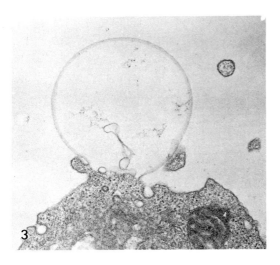

FIG. 3. Transmission electron micrograph of a blister on the surface of a lymphocyte fixed in 2.5% glutaraldehyde for 30 min and postfixed in 1% OsO_4 for 5 min at room temperature. ×27,000.

REFERENCES

1. Hasty, D. L., and Hay, E. D. (1978): Freeze-fracture studies of the developing cell surface. II. Particle-free membrane blisters on glutaraldehyde fixed corneal fibroblasts are artifacts. *J. Cell Biol.,* 78:756–768.
2. Scott, R. E. (1976): Plasma membrane vesiculation: A new technique for isolation of plasma membranes. *Science,* 194:743–745.
3. Shelton, E., and Mowczko, W. E. (1977): Membrane blebs: A fixation artifact. *J. Cell. Biol.,* 72:206a.
4. Shelton, E., and Mowczko, W. E. (1978): Membrane blisters: A fixation artifact. A study in fixation for scanning electron microscopy. *Scanning,* 1:166–173.

Freeze Fracture: Methods, Artifacts, and Interpretations, edited by J. E. Rash and C. S. Hudson. Raven Press, New York © 1979.

Subplasmalemmal Vesicle Clusters: Real or Artifact?

Karl H. Pfenninger

Department of Anatomy, Columbia University, New York, New York 10032

Ever since the first ultrastructural study of nerve growth cones investigators have been puzzled by the occurrence of clusters of large clear vesicles underlying the plasmalemma (3,4,8,12,24,26). I will refer to these vesicles as "subplasmalemmal vesicles" (SPVs) for the time being, until the elucidation of their function enables us to find a more meaningful name. Because of the great variability in the preservation of SPV clusters, it has been difficult to determine whether they are real or artifact (c.f., 25). The same problem has arisen in other cellular systems where similar vesicular structures and preservation problems have been encountered (9,10,22).

The best way to determine if a given structure exists in the living cell is to observe it directly in the living state by using phase or interference contrast microscopy. Unfortunately, this method is limited to the range of light microscopic resolution. Beneath it, electron optical analysis of chemically or physically fixed material must be used. Therefore, a kind of uncertainty principle applies: one cannot ever be entirely sure of the appearance of reality at the ultrastructural level. The question of whether a particular structure is physiological or artifactual cannot be answered simply by applying an "ideal" (chemical) fixative; one's choice is bound to be the result of a subjective decision. Rapid freezing, a physical mode of fixation which is not subject to chemically induced artifacts, may offer a better solution (6,11,13). Unfortunately, this technique, so far, has not been applied successfully to nerve growth cones *in vivo* or *in vitro*. One problem has been the delicacy of growing neurites and their tips. Furthermore, the nonmetallic culture substratum (that one is forced to use in nerve tissue culture) on the one hand, and the fluid layer (culture medium) above the neurites *in vitro* on the other hand, slow down heat withdrawal from either side enough so that ultrastructure is distorted by ice crystal formation. In addition, if a given structure is delicate, it might be distorted *despite* rapid freezing during the subsequent processing of tissue for electron microscopy. Thus, damage could conceivably be introduced by the techniques of freeze-substitution or freeze-drying. For growing neurites *in vivo*, the situation is further complicated by the fact that one would not be likely to find the relatively rare growth cones undamaged within the quick-frozen (vitrified) part of a tissue block, a superficial layer about 5 μm thick.

The thin nerve growth cones, if grown *in vitro,* are amenable to detailed light microscopic analysis (c.f., 20). Therefore, one can attempt to correlate structures in live cells discernible in the light microscope with ultrastructural features visible at the electron microscopic level. Furthermore, the accessibility of the surface of neurites growing in culture permits a wide range of experiments for the characterization of the properties and movements of their membranes and membrane components. Such experiments can be of significance for our understanding of cellular ultrastructure and complement description of the static fixed or frozen state. The following text is a summary of various studies on membranes of growing neurites which pertain to the question of the nature of SPVs.

RESULTS AND INTERPRETATIONS

As observed by Pomerat et al. (20), nerve growth cones contain a variable number of highly refractile, vacuole-like structures. During neuritic growth these "vacuoles" tend to increase slowly in size for several minutes and may then be seen to collapse or disappear *within seconds.* Over longer periods of observation, they may also change their location within the growth cone. If the growth cones are studied during the infusion of a fixative into the culture medium, persistence of the "vacuoles" is observed but no new "vacuoles" are formed. Such fixed specimens can then be processed for serial thin sectioning and ultrastructural analysis. As shown in Figs. 1 and 2, the light microscopically identified "vacuoles" appear to consist of clusters of large, clear vesicles: SPVs. The observation that "vacuoles" seen with the light microscope in *live,* growing neurites can be identified electron microscopically with clusters of SPVs indicates that these structures correspond to a physiological entity. Even if we cannot know how they looked while still alive, they must either be, or be formed from, a real structure.

A wide variety of primary fixatives may be used to preserve SPVs for ultrastructural analysis. These include glutaraldehyde, acrolein, osmium tetroxide, and combinations thereof. The fixatives may be applied in various buffers and ionic environments. Furthermore, the reagents may be added to the cell cultures either slowly or rapidly and at various temperatures from $-2°C$ to $+36°C$. Although the quality of preservation, especially the degree of swelling, varies under these different conditions, SPV clusters can be observed consistently in the nerve growth cones (Figs. 3–5). Prior reports (9) that such vesicle clusters cannot be found after fixation with a combination of glutaraldehyde and osmium tetroxide do not apply to growing neurites (Fig. 3). This discrepancy in results may have to do with the fact that, in cell cultures, the action of fixatives is particularly rapid and, therefore, different from that in tissue blocks. However, the free accessibility of cell surfaces in culture also makes them particularly vulnerable. This may be why gradual exposure of cultured neurons to a fixative,

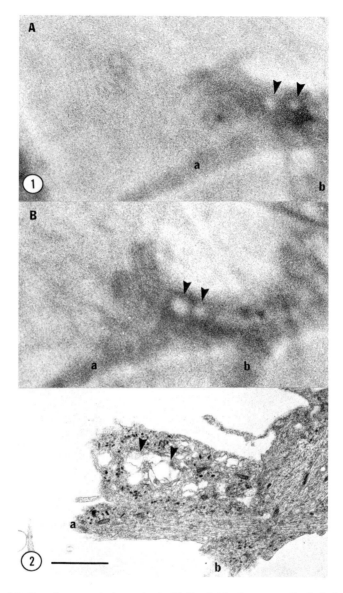

FIGS. 1 and 2. Growth cone photographed with the light microscope 2 min before **(1A)** and 7 min after **(1B)** onset of aldehyde fixation. Note the persistence of two large, refractile "vacuoles" (arrowheads). The same growth cone was then serially sectioned and studied with the electron microscope (Fig. 2). The "vacuoles" seen light microscopically can be identified with clusters of vesicles (arrowheads). Because of the constraints on the mode of application of the fixative due to continuous light microscopic monitoring, the fixation of this specimen was not ideal (c.f., Fig. 3) and coalescence of some vesicles has occurred. a and b, corresponding structures in light and electron micrographs. Culture of rat superior cervical ganglion, 3 days *in vitro*. Calibration bar, 2 μm.

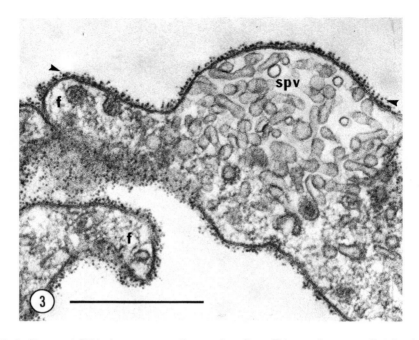

FIG. 3. Cluster of SPV of a nerve growth cone in culture. This specimen was first fixed with glutaraldehyde, washed with glycine-containing buffer, treated with neuraminidase to cleave off terminal sialic acid residues, and finally labeled with ferritin-conjugated agglutinin I from *Ricinus communis* (specific for galactosyl residues). Then, the tissue was processed for electron microscopic analysis (c.f. 18). Note the absence of swelling of SPVs and the uniform layer of ferritin-lectin label all over the cellular processes, including the plasmalemma covering the SPV cluster (arrow heads). f, growth cone filopodia. Rat spinal cord, grown for 5 days *in vitro*. Calibration bar, 0.5 μm.

a fairly unorthodox procedure, is successful in preserving nerve growth cones and ·SPVs. In our hands, the best structural preservation of growing neurons is obtained by slow infusion into the culture medium of a fixative containing 1.5% glutaraldehyde in 0.1 M phosphate buffer, pH 7.3, with 0.4 mM calcium chloride and 120 mM glucose added. Further processing of the samples, although somewhat less critical, must be very gentle, involving the gradual change of reagent solutions at each step. An example of a well-preserved SPV cluster is shown in Fig. 3 (this preparation was labeled with a lectin-ferritin conjugate following fixation).

Freeze-fracturing of nerve growth cones, after careful fixation with aldehyde and glycerol treatment, reveals plasmalemmal protrusions filled with vesicles corresponding in size, morphology and location to the SPV clusters seen in thin sections (Fig. 6). Analogous structures are also seen in supporting cells *in vivo* and *in vitro* (Figs. 7 and 8). In all cases, SPVs are found to be essentially free of intramembranous particles (IMPs). Furthermore, the plasmalemma cover-

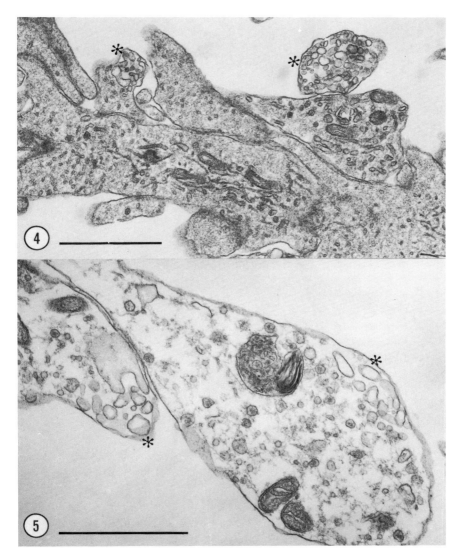

FIGS. 4 and 5. Nerve growth cones fixed with a combination of acrolein with glutaraldehyde (Fig. 4) and of OsO_4 with glutaraldehyde (Fig. 5), respectively. Although not ideally preserved, subplasmalemmal vesicle clusters are evident in both preparations (asterisks). 3-day-old cultures of rat superior cervical ganglion. Calibration, 1 μm in both figures.

ing these vesicle clusters is largely IMP-free, a situation which is particularly clear in glial cells, which exhibit in the plasmalemma surrounding the vesicle mounds a higher density of IMPs than the nerve growth cones (Fig. 6 and 7; see also refs. 14,15). It has been reported that freeze-fracture does not reveal

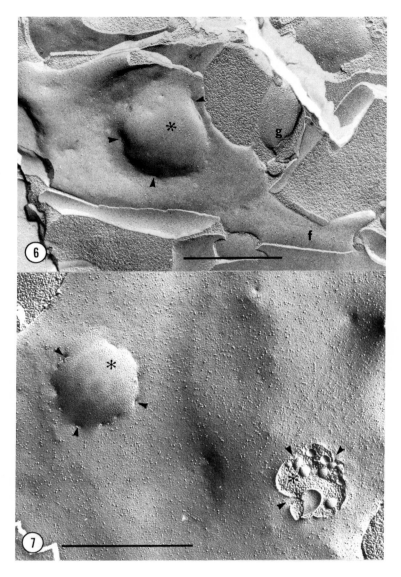

FIGS. 6 and 7. "Mounds" (asterisks) filled with subplasmalemmal vesicles as seen by freeze fracture in a nerve growth cone (Fig. 6) and a supporting cell, probably glia (Fig. 7). The protoplasmic leaflet of the plasmalemma is exposed in both cases. Note the paucity of intramembranous particles in the plasmalemma covering the SPV clusters, and the pits (arrow heads) which are the openings of basal cisternae (see text). The minute "cobblestone" effect in Fig. 7 probably results from slight contamination from frozen water vapor (see refs. 21 and 23). One SPV cluster is cross-fractured exhibiting particle-free vesicles (Fig. 7, lower right). f, growth cone filopodium, particle poor; g, glial process (protoplasmic leaflet), richer in particles. Cultures of rat spinal cord and dorsal root ganglion, 50 days *in vitro* (Fig. 6), and of rat olfactory bulb, 11 days *in vitro* (Fig. 7). Shadowing direction, from bottom (Fig. 6) and from lower left (Fig. 7). Calibration, 1 μm in both figures.

such vesicle mounds in the absence of aldehyde fixation (9). This may be because the investigators worked on a different system or because these delicate structures collapse or are destroyed when frozen without prior stabilization or cryoprotection. Our freeze-fracture studies also demonstrate that the structure of SPV clustering sites is even more complex than may appear during examination of thin sections. At the base and lateral margin of the SPV cluster, there are plasmalemmal invaginations which form, just beneath the SPVs, a branched network of cisternae which remain open to the extracellular space (4,26). By contrast to that of SPVs, the membrane of these cisternae contains as many IMPs as the plasma membrane surrounding the SPV cluster area. It is evident that SPV mounds are highly organized structures.

Despite the structural similarity of mound plasmalemma and SPV membranes, i.e., the virtual lack of IMPs in both membrane types, SPVs are not generated by artifactual vesiculation of the plasmalemma. This conclusion can be drawn from the following experiment: neurons sprouting in culture were fixed in the presence of a high concentration (10 mg/ml) of an extracellular tracer such as ferritin. Ultrastructural analysis showed that SPVs of well-fixed, i.e., not or only minimally ballooned, clusters did not contain the tracer (Fig. 9). This result excludes the possibility of physiological (by pinocytosis) or artifactual (by vesiculation) intake of extracellular tracer into SPVs, and is further supported by membrane labeling studies with ferritin-conjugated lectins reported elsewhere (16–18). The apparent intake of tracer into SPVs described by Birks et al. (2), delCerro (7), and Bunge (5) has not been shown to be time-dependent and, therefore, may be the result of *fixation-induced* artifactual fusion of SPVs with neighboring tracer-containing compartments, such as the basal cisternae that lie next to mound SPVs.

The absence of IMPs in the plasmalemma covering SPV clusters has been interpreted to be the result of extrusion of integral membrane components due to incomplete stabilization of the membrane with aldehyde (9,10). However, when growing neurites were fixed with aldehyde, washed with glycine-containing buffer to quench the remaining aldehyde groups, and then labeled with ferritin-lectin conjugates, the surfaces of SPV clusters were as densely labeled with the carbohydrate marker as surrounding plasmalemmal areas (Figure 3; see also ref. 18). This result rules out the possibility that, in this case, aldehyde fixation induces such major changes in membrane architecture as the lateral displacement of integral membrane glycoproteins and glycolipids.

If lectin-labeling is carried out in the form of pulse-chase experiments, it can be shown that the surface of an SPV cluster is the site where new lectin receptors appear selectively during neuritic growth and plasmalemmal expansion (14,17–19). This finding greatly stresses the physiological significance of SPVs—*whatever their actual appearance* in the live state. This last point indicates that the argument over the structural appearance of SPV clusters may be of only limited importance. Because their occurrence is reproducible in place and in relation to the point of insertion of new lectin receptors into the plasma membrane, they must reflect a structurally and functionally unique point in the

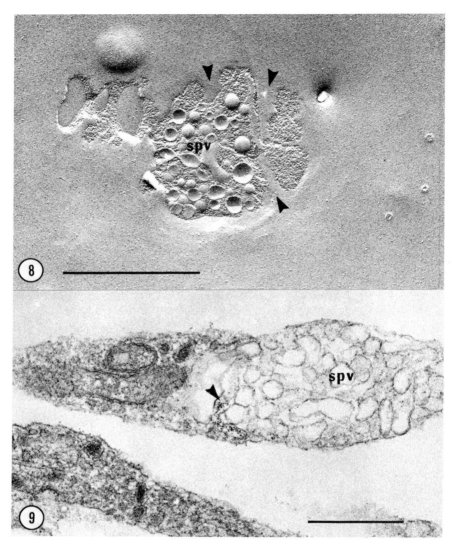

FIG. 8. Cluster of SPV seen from a vantage point inside the cell (the external plasmalemmal leaflet is exposed). Note the virtual absence of particles in the membrane of SPVs, and the basal invaginations of the plasmalemma (arrow heads). Supporting cell in a culture of rat spinal cord, grown *in vitro* for 7 days. Shadowing direction, from bottom. Calibration, 1 μm.

FIG. 9. Growing neurites in a culture of rat superior cervical ganglion (4 days *in vitro*) with a large cluster of SPV. This specimen was first exposed for 1 min to the extracellular tracer ferritin (10 mg/ml) and then glutaraldehyde-fixed and processed for electron microscopic examination. Note the absence of the tracer in SPVs. The only structure labeled in this picture is a basal cisterna (arrow head). Calibration, 0.5 μm.

cellular membrane system. Therefore, their significance is undiminished even if the configuration the SPV clusters show in aldehyde-fixed material does turn out to be artifactual.

A further argument for the physiological significance of SPVs is their temporally selective occurrence in a wide variety of eukaryotic cells: SPVs have only been observed in growing systems, such as developing or regenerating tissues and cells proliferating or sprouting *in vitro*. Furthermore, they seem to occur in specific locations, such as the area just proximal to the leading edge of expanding glial and fibroblast processes, the cleavage furrow of dividing zygotes (1) and, as described here, the growth cone. Interestingly, all these areas are putative regions of membrane growth.

CONCLUSIONS

The various descriptive and experimental studies summarized here establish that the clear, large, somewhat pleomorphic vesicles (SPVs), which are frequently found in nerve growth cones, represent physiological structures. We cannot be certain that the ultrastructure of SPV clusters described here is *identical* to that in the live cell. Yet, certain statements can be made: SPVs are characterized by a lack of IMPs as is the plasmalemma covering their clusters. The particle-free plasmalemma is rich in a variety of glycoconjugates, as demonstrated by lectin-binding experiments. Studies carried out in this laboratory actually suggest that SPVs contain the membrane precursor for plasmalemmal growth (14,18,19). On the basis of the data presently available, it is not possible to determine to what extent the SPV clusters and IMP-free plasmalemmal regions described here are related to those described by other authors in other cell systems (10,22). However, comparison of the various results clearly indicates (a) that a judgment on the actual appearance in life of SPVs, or of any structure too small to be seen live with the light microscope, on the basis of different fixation or freezing procedures is impossible, and (b) that broad generalizations from one cellular system to another are not adequate. Ever since the freeze-fracture technique was developed investigators have been able to observe artifactual displacement of intramembranous particles before freezing in a variety of cellular systems. Yet, the occurrence of such an artifact does not preclude by any means the possibility that, under certain *physiological* conditions, the cell might alter or modify its plasmalemma in specific regions with a morphologically similar result. In such situations, the distinction between artifact and physiologic appearance will always be particularly difficult. At the same time, this distinction should not permit critical cellular events to be overlooked.

ACKNOWLEDGMENTS

Supported by U.S. Public Health Service grant NS-13466, and grant BNS-18513 from the National Science Foundation and an I.T. Hirschl Career Scientist Award.

REFERENCES

1. Arnold, J. M. (1976): Cytokinesis in animal cells: new answers to old questions. In: *The Cell Surface in Animal Embryogenesis and Development,* edited by G. Poste and G. L. Nicolson, pp. 55–80. North Holland, Amsterdam.
2. Birks, R. I., Mackey, M. C., and Weldon, P. R. (1972): Organelle formation from pinocytotic elements in neurites of cultured sympathetic ganglia. *J. Neurocytol.,* 1:311–140.
3. Bodian, D. (1966): Development of fine structure of spinal cord in monkey fetuses. I. Motoneuron neuropil at time of onset of reflex activity. *John Hopkins Med. J.,* 119:129–149.
4. Bunge, M. B. (1973): Fine structure of nerve fibers and growth cones of isolated sympathetic neurons in culture. *J. Cell Biol.,* 56:713–735.
5. Bunge, M. B. (1977): Initial endocytosis of peroxidase or ferritin by growth cones of cultured nerve cells. *J. Neurocytol.,* 6:407–439.
6. Chandler, D. E. Quick freezing avoids specimen preparation artifacts in membrane fusion studies. *(this volume).*
7. delCerro, M. P. (1974): Uptake of tracer proteins in the developing cerebellum particularly by the growth cones and blood vessels. *J. Comp. Neurol.,* 157:245–280.
8. delCerro, M. P., and Snider, R. S. (1968): Studies on the developing cerebellum. Ultrastructure of the growth cones. *J. Comp. Neurol.,* 133:341–362.
9. Hasty, D. L., and Hay, E. D. (1978): Freeze-fracture studies of the developing cell surface. II. Particle-free membrane blisters on glutaraldehyde-fixed corneal fibroblasts are artifacts. *J. Cell Biol.,* 78:756–768.
10. Hay, E. D., and Hasty, D. L. Extrusion of particle-free membrane blisters during glutaraldehyde fixation. *(this volume).*
11. Heuser, J. E., Reese, T. S., and Landis, D. M. D. (1976): Preservation of synaptic structure by rapid freezing. *Cold Spring Harbor Symp. Quant. Biol.* XL:17–24.
12. Kawana, E., Sandri, C., and Akert, K. (1971): Ultrastructure of growth cones in the cerebral cortex of the neonatal rat and cat. *Z. Zellforsch.,* 115:284–298.
13. Ornberg, R. L., and Reese, T. S. Freezing artifacts in *Limulus* amebocytes. *(this volume).*
14. Pfenninger, K. H. (1979): Synaptic Membrane Differentiation. In: The *Neurosciences: Fourth Study Program,* edited by F. O. Schmitt, pp. 779–795. MIT Press, Cambridge, Mass.
15. Pfenninger, K. H., and Bunge, R. P. (1974): Freeze-fracturing of nerve growth cones and young fibers. A study of developing plasma membrane. *J. Cell Biol.* 63:180–196.
16. Pfenninger, K. H., and Maylié-Pfenninger, M.-F. (1975): Distribution and fate of lectin binding sites on the surface of growing neuronal processes. *J. Cell Biol.* 67:322a.
17. Pfenninger, K. H., and Maylié-Pfenninger, M.-F. (1977): Localized appearance of new lectin receptors on the surface of growing neurites. *J. Cell Biol.,* 75:54a.
18. Pfenninger, K. H., and Maylié-Pfenninger, M.-F. (1978): Characterization, distribution, and appearance of surface carbohydrates on growing neurites. In: *Neuronal Information Transfer,* edited by A. Karlin, H. Vogel, and V. M. Tennyson, pp. 373–386. Academic Press, Inc., New York.
19. Pfenninger, K. H., Small, R., and Maylié-Pfenninger, M.-F. (1979): Insertion of membrane components during plasmalemmal growth in neurons. *Proc. Cold Spring Harbor Symp. Membrane Biogenesis.* p. 96. Cold Spring Harbor Laboratory, Cold Spring Harbor, New York.
20. Pomerat, C. M., Hendelman, W. J., Raiborn Jr., C. W., and Massey, J. F. (1967): Dynamic activities of nervous tissue in vitro. In: *The Neuron,* edited by H. Hyden, pp. 119–178. American Elsevier Publishing Co., Inc., New York.
21. Rash, J. E., Graham, W. F., and Hudson, C. S. Sources and rates of contamination in a conventional Balzers freeze-etch device. *(this volume).*
22. Shelton, E., and Mowczko, W. E. Scanning electron microscopy of membrane blisters produced by glutaraldehyde fixation and stabilized by post-fixation in osmium tetroxide. *(this volume).*
23. Steere, R. L., Erbe, E. F., and Moseley, J. M. Controlled contamination of freeze-fractured specimens. *(this volume).*
24. Tennyson, V. M. (1970): The fine structure of the axon and growth cone of the dorsal root neuroblast of the rabbit embryo. *J. Cell Biol.,* 44:62–79.
25. Wessels, N. K., Nuttall, R. P., Wrenn, J. T., and Johnson, S. (1976): Differential labeling of the cell surface of single ciliary ganglion neurons in vitro. *Proc. Natl. Acad. Sci. USA,* 73:4100–4104.
26. Yamada, K. M., Spooner, B. S., and Wessels, N. K. (1971): Ultrastructure and function of growth cones and axons of cultured nerve cells. *J. Cell Biol.,* 49:614–635.

Freeze Fracture: Methods, Artifacts, and Interpretations, edited by J. E. Rash and C. S. Hudson. Raven Press, New York © 1979.

Quick Freezing Avoids Specimen Preparation Artifacts in Membrane-Fusion Studies

Douglas E. Chandler

Department of Physiology, University of California, San Francisco, California 94143

Freeze fracture exposes broad panoramas of membrane interior making this technique particularly well suited for studying membrane fusion. Fusion of secretory granule and plasma membranes during exocytosis has already been studied in a number of cells (1,2,5–9) with the result that several intermediate stages in the fusion process have been visualized and interpreted. However, it has not been generally appreciated that membrane events as rapid as fusion may be altered by the conventional methods used to prepare tissue for freeze fracture. In fact, fixation with aldehydes, followed by cryoprotection with glycerol and freezing in Freon 22 can induce membrane fusion as well as formation of "intermediates" that are artifacts.

We have been able to evaluate these artifacts by using an entirely different method of stopping cell processes, that is, extremely quick freezing. Quick freezing is accomplished by the method of Heuser et al. (3); cells are placed at the tip of a vertical plunger that drops by gravity onto a copper block cooled to 4°K by liquid helium. Tissue frozen in this way can be freeze fractured directly without exposure to chemical fixatives or cryoprotectants.

Recently, we used this technique to study exocytosis of cortical granules in sea urchin eggs. Cortical granules lie in a single layer just below the plasma membrane of the egg (Fig. 1) and, at fertilization, these undergo fusion with the plasma membrane. Exocytosis starts at the point of sperm attachment and spreads as a wave over the entire surface of the egg.

We found what at first appeared to be "intermediates" in exocytosis by fixing eggs for 1 hr with glutaraldehyde [1.8% in artificial seawater (ASW)], then glycerinating for 1.5 hr (30% glycerol in ASW), and freezing in Freon 22. This resulted in numerous groups of etchable pores (Fig. 2), which, in appropriate fractures, were seen to be spot fusions between granule and plasma membranes (Fig. 3). Thus each group of pores represented multiple openings into an underlying granule. Some pores had a diaphragm of membrane free of intramembrane particles (IMP) that was continuous with both the plasma membrane and the granule membrane (*arrow,* Fig. 3). This diaphragm appeared to be a single bilayer judging from the fact that there was no jump in the fracture plane from one membrane to the other at the diaphragm. Although occasional fusions between granule and plasma membranes were seen in unfertilized eggs, groups

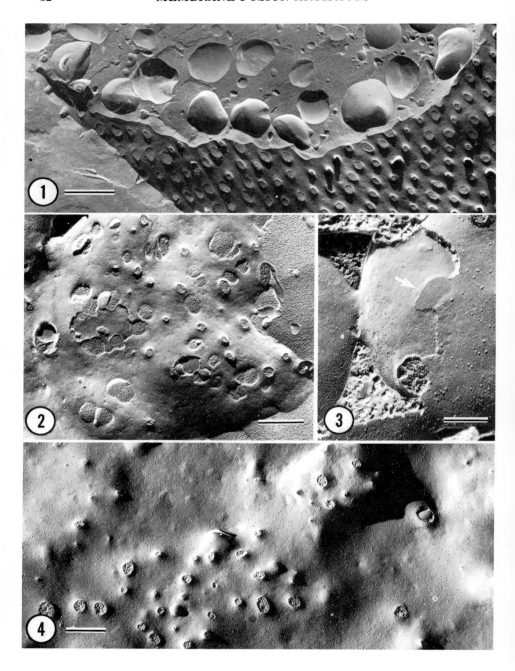

of pores leading to a single granule occurred only during cortical granule exocytosis.

In contrast, when eggs were "quick frozen" without fixation or glycerination, these intermediates were not found (data to be published elsewhere). At the front of the exocytosis wave, depressions marked the sites where single granules had fused, and these were well separated by plasma membrane that appeared unchanged from that in unfertilized eggs and exhibited no pores. Thus granules looked either intact or wide open to the outside through a single orifice suggesting that the true fusion intermediates must be extremely short lived. This discrepancy in results between fixed and quick-frozen tissue led us to believe that the pores and membrane diaphragms observed were artifacts of fixation, glycerination, or method of freezing.

The cause of the artifact became clear when eggs were fixed and then quick frozen either with or without being glycerinated. Eggs fixed and quick frozen without glycerination exhibited no pores (Fig. 4), whereas those quick frozen after glycerination exhibited multiple pores identical to those seen in glutaraldehyde-fixed eggs frozen in Freon 22 (Fig. 2). This indicated that the pores result from fusion of granule and plasma membranes during glycerination of fixed tissue. Their abundance depended on the concentration of glycerol; 10% glycerol produced only a few pores, whereas 20 or 30% produced numerous groups of pores.

Glycerol-induced membrane fusion also occurred between membranes of adjacent cortical granules in both fertilized and unfertilized eggs. Frequently, the area of granule–granule contact contained an IMP-free membrane diaphragm that was continuous with membranes of both granules (Fig. 5). At higher magnification one could see channels of cytoplasm between diaphragms where the two granule membranes had not yet fused (arrow, Fig. 6).

Artifactual fusion of membranes during tissue preparation was not limited to the sea urchin egg. We frequently found fusions between granules in unstimulated mast cells that were fixed, glycerinated, and frozen in Freon (Fig. 7). Yet, if mast cells were quick frozen without fixation or glycerination, no fusions

---

FIGS. 1–4. 1: Cortex of an egg from the sea urchin *S. Purpuratus.* The cortical granules lie in a single layer just beneath the cell surface, yet they are distinctly separate from each other and the plasma membrane. Specimen was quick frozen 20 sec after insemination when exocytosis had not yet occurred. ×13,000; marker = 1 μm. **2:** P-face of the plasma membrane of an egg fixed 30 sec postinsemination, then glycerinated in 30% glycerol and quick frozen. Numerous groups of etchable and nonetchable pores can be seen in this area where granule exocytosis is underway. Identical results were obtained whether eggs were quick frozen or frozen in resolidifying Freon. ×24,000; marker = 0.5 μm. **3:** Fractures that jumped from plasma to granule membrane revealed that pores were spot fusions between these two membranes. IMP-free diaphragms covering some pores were continuous with both granule and plasma membrane *(arrow).* Specimen was fixed, glycerinated, and frozen as in Fig. 2. ×56,000; marker = 0.25 μm. **4:** Plasma membrane of an egg fixed 30 sec postinsemination, then quick frozen. Exocytosis is well underway as indicated by presence of microvilli-free patches of fused granule membrane. No artifactual pores are seen because these cells were not glycerinated. ×21,000; marker = 0.5 μm.

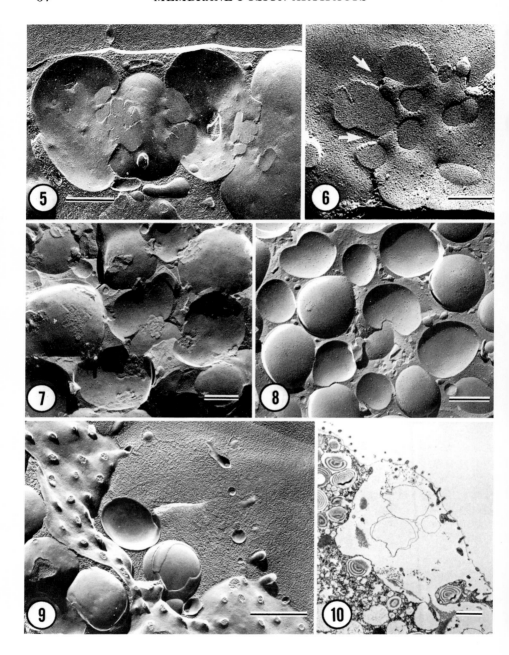

were seen; the fracture plane jumped cleanly between membranes of adjacent granules (Fig. 8). Fusion of granule membranes was also absent in thin sections of glutaraldehyde-fixed mast cells suggesting that here too glycerination was required for fusion to occur.

Comparison of quick-frozen eggs and glutaraldehyde-fixed eggs also indicated that artifactual changes in membrane structure occur during fixation as well as glycerination. Depressions, resulting from fusion of cortical granules with the plasma membrane, routinely contained lipid vesicles ranging from about 1 to 5 μm in diameter and having few or no IMPs (Fig. 9). Transmission micrographs showed that the vesicles were composed of typical unit membranes, and had single or multiple walls (Fig. 10). Vesicles of this type were found in glutaraldehyde-fixed eggs whether or not they were subsequently glycerinated. However, such vesicles were consistently absent in tissue that had been quick frozen without fixation, suggesting that aldehyde fixation had either provoked (or failed to prevent) budding off of cell membranes undergoing rapid structural changes during fusion. Unstable and "unfixable" membrane appeared to be present only during exocytosis; vesicles were not seen before cortical granule release, and once secretion was complete the lipid vesicle artifacts disappeared abruptly.

It is apparent from these results that the early events in membrane fusion are difficult to observe and retain in their physiological form. Glutaraldehyde fixation appears to be insufficient for this purpose because structurally unstable membranes can either slough off during fixation to form vesicles or can be fused by glycerol following fixation. Furthermore, the artifacts produced look remarkably similar to fusion intermediates already described in other cell types. Clearing of IMPs in the plasma membrane has been seen during secretion from mast cells (1,5) and β-cells in the pancreas (6), during the acrosomal reaction in sperm (2), and during apocrine secretion of milk from the mammary gland (7); formation of small, IMP-free vesicles has been observed at sites of myoblast fusion (4); and finally, formation of IMP-free, single bilayer diaphragms, shared

---

FIGS. 5–10. 5: Fusion between adjacent cortical granules resulting in formation of intramembrane particle-free membrane diaphragms. Specimen was fixed, glycerinated, and frozen as in Fig. 2. ×25,000; marker = 0.5 μm. **6:** Fusion between two adjacent cortical granules showing diaphragms shared by both granule membranes. Between diaphragms are cytoplasmic channels where granule membranes have not yet fused *(arrows)*. Specimen was fixed, glycerinated, and frozen as described in Fig. 2. ×58,000; marker = 0.25 μm. **7:** Periphery of an unstimulated mast cell showing numerous fusions between adjacent granules. Specimen was fixed in glutaraldehyde (2% in 0.1 M cacodylate buffer, pH 7.4 for 1 hr), glycerinated (22% glycerol:78% 0.1 M cacodylate buffer), and frozen in Freon 22. ×20,000; marker = 0.5 μm. **8:** Interior of a quick-frozen mast cell. Note there are no fusions between membranes of adjacent granules; the fracture plane jumps cleanly from one granule to the next. ×21,000; marker = 0.5 μm. **9:** Lipid vesicles fill the depression where cortical granules have fused with the plasma membrane. Specimen was fixed, glycerinated, and frozen as described in Fig. 2, except that 10% glycerol was used during glycerination. ×15,000; marker = 1 μm. **10:** Thin section of an egg fixed 30 sec postinsemination showing an exocytotic depression filled with single- and double-walled, lipid bilayer vesicles. ×7,000; marker = 1 μm.

by granule and plasma membrane, occurs during exocytosis in zoospores (8). This similarity warrants use of caution in interpreting fusion intermediates seen in tissue fixed with glutaraldehyde and glycerinated.

It would be presumptuous, at this point, to suggest that all fusion intermediates previously documented are artifacts. There is a great deal of variability in the quality of fixation from one tissue to the next or even between different membranes in the same cell, and it is the quality of fixation that probably determines whether artifacts occur. For example, in the unstimulated mast cell, artifactual fusions were never seen between plasma and granule membranes even in cells where numerous fusions had occurred between membranes of adjacent granules. In addition, there may be some strength in the argument that fusion intermediates previously studied are not artifacts because they have been seen in both thin sections and freeze fracture. Pentalaminar-like contacts (1) and clearing of surface ligands (5), as seen in thin sections, have been related to IMP clearings in the plasma membrane of mast cells during secretion, and the single bilayer diaphragms that form between granule and plasma membrane during zoospore exocytosis, can be seen by both methods (8).

Yet, this report is a clear warning that artifact production is possible and likely in fixed tissues, and that we may only have begun to understand what artifacts are involved in preparation of tissues for either embedding or freeze fracture. What is needed is to use alternative methods of tissue preservation, such as quick freezing, to study membrane fusion, evaluate artifacts, and to determine whether artifact production itself can teach us something about how biological membranes fuse.

SUMMARY

Cortical granule exocytosis in eggs of the sea urchin *Stronglyocentrotus purpuratus* was studied by using quick-freezing or conventional fixation and glycerination techniques to prepare tissue for freeze fracture. Glycerination after fixation induced artifactual fusion between granule and plasma membranes and between membranes of adjacent granules. Some fusion sites retained single bilayer diaphragms that were clear of intramembrane particles and continuous with both granule and plasma membranes. This was not seen in tissue that was quick frozen and it is suggested that this method avoids artifacts that look remarkably similar to intermediates reported in previous membrane fusion studies.

ACKNOWLEDGMENTS

I thank Dr. John Heuser for advice and use of his laboratory throughout this study. The work was carried out during tenure of a postdoctoral fellowship from the Pharmaceutical Manufacturer's Association Foundation and was supported by grants to Dr. Heuser from the U.S.P.H.S. (NS 11979) and the Muscular Dystrophy Association.

REFERENCES

1. Chi, E. Y., Lagunoff, D., and Koehler, J. K. (1976): Freeze-fracture study of mast cell secretion. *Proc. Natl. Acad. Sci. USA,* 73:2823–2827.
2. Friend, D. S., Orci, L., Perrelet, A., and Yanagimachi, R. (1977): Membrane particle changes attending the acrosome reaction in guinea pig spermatozoa. *J. Cell Biol.,* 74:561–577.
3. Heuser, J. E., Reese, T. S., and Landis, D. M. D. (1976): Preservation of synaptic structure by rapid freezing. *Cold Spring Harbor Symp. Quant. Biol.,* 40:17–24.
4. Kalderon, N., and Gilula, N. B. (1977): Analysis of the events involved in myoblast fusion. *J. Cell Biol.,* 75:222a.
5. Lawson, D., Raff, M. C., Gomperts, B., Fewtrell, C., and Gilula, N. B. (1977): Molecular events during membrane fusion. A study of exocytosis in rat peritoneal mast cells. *J. Cell Biol.,* 72:242–259.
6. Orci, L., Perrelet, A., and Friend, D. S. (1977): Freeze-fracture of membrane fusions during exocytosis in pancreatic β-cells. *J. Cell Biol.,* 75:23–30.
7. Peixoto de Menezes, A., and Pinto da Silva, P. (1978): Freeze-fracture observations of the lactating rat mammary gland. *J. Cell Biol.,* 76:767–778.
8. Pinto da Silva, P., and Nogueira, M. L. (1977): Membrane fusion during secretion. A hypothesis based on electron microscope observation of *Phytophthora palmivora* zoospores during encystment. *J. Cell Biol.,* 73:161–181.
9. Satir, B., Schooley, C., and Satir, P. (1973): Membrane fusion in a model system. Mucocyst secretion in Tetrahymina. *J. Cell Biol.,* 56:153–176.

Freeze Fracture: Methods, Artifacts, and Interpretations, edited by J. E. Rash and C. S. Hudson. Raven Press, New York © 1979.

Artifacts of Freezing in *Limulus* Amebocytes

R. L. Ornberg and T. S. Reese

Section on Functional Neuroanatomy, Laboratory of Neuropathology and Neuroanatomical Sciences, National Institute of Neurological and Communicative Disorders and Stroke, National Institutes of Health, Bethesda, Maryland 20205

This paper illustrates artifacts of freezing we have encountered in *Limulus* amebocytes (11). No attempt is made to examine theory of freezing or ice crystal damage, which has been extensively covered elsewhere (9,10,15). *Limulus* amebocytes are frozen directly by pushing a drop of freshly drawn blood (20 μl) against a copper block cooled close to the temperature of liquid helium (11,7,8). Fixatives and cryoprotectants are not necessary with this method of freezing.

This procedure is assumed to result in a planar freezing front, where the freezing starts along the contact with the copper block and proceeds inward. The freezing with *Limulus* blood has proven to be rather "good" within several micrometers of the initial contact surface, perhaps aided by the large concentrations of electrolytes (900 mM) and proteins in the blood of these animals. However, a gradient in the size of the ice crystals is present and this gradient is particularly easy to see in this preparation because the blood of the horseshoe crab contains a large protein, hemocyanin (8×10^5 D), which is displaced by ice crystals and therefore outlines them.

The presence of an extracellular marker for ice crystal size means that crystals can also be measured in freeze-substituted material, cut exactly perpendicular to the freezing surface (4). The availability of views perpendicular to the freezing surface means that the diameters of ice crystals, as inferred from the patterns of hemocyanin distribution, could be compared to the distance from the initial plane of freezing (Fig. 1). These values are scaled along the top and vertical axes of Fig. 2, which also shows the depth in a block of liver frozen at different times after contact with the freezing surface (7). Because the depth below the tissue surface and ice crystal size are both linear dimensions, any uniform shrinkage in the embedded tissue should not affect this relationship, at least for our purposes.

Figure 2 was used as a calibration curve for frozen drops of amebocyte blood fractured with a razor blade at $-110°C$ in a Balzers 360 M. In this apparatus, the plane of fracture is parallel to the initial freezing surface. With the help of the calibration curve, an estimate of the distance of a cell from this surface could be based on the sizes of the ice crystals surrounding it; these distances

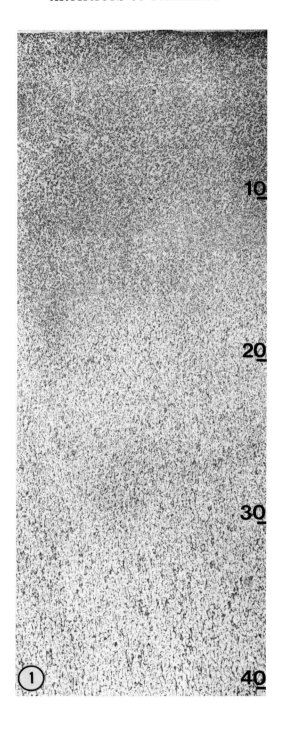

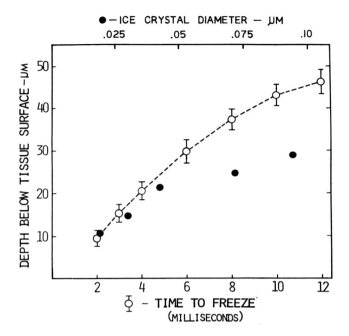

FIG. 2. Depths in a block of liver reached by freezing at different times after contact with the cold block, as measured with a capacitance circuit (open circles filled by dotted line; redrawn from 7). The time to freeze represents a maximum, because it actually corresponds to the time to cool the ice to $-80°C$. We superimposed estimates of ice crystal size at different depths in drops of freeze-substituted *Limulus* blood (black dots, scaled along top). This result means that the *extracellular* ice crystals surrounding amebocytes are approximately 30 nm in diameter at a depth of 15 μm below the freezing surface, where the tissue has frozen in less than 3 msec.

are indicated for each illustration. Within this range of distances, displacements of hemocyanin by ice crystals grew by approximately 3 nm/μm of depth in the tissue. This is a somewhat better result than that reported where liquid nitrogen rather than liquid helium is used to cool the copper cold block (4,5; see Appendix I in 7).

The plasma membrane of amebocytes showed a regular series of changes as the distance from the freezing surface increased (Fig. 3). Within 10–15 μm of the surface (Figs. 4 and 6), fracture faces were smooth and marked by sharply defined intramembrane particles. Deeper, both fracture faces acquired a cobblestone appearance and the size of these membrane deformations progressively

FIG. 1. Ice crystal gradient in first 40 μm of freeze-substituted *Limulus* blood next to freezing surface (top). Depths in the tissue in micrometers are indicated at right. Hemocyanin molecules (smallest visible specks) begin to be displaced from their normal distribution at approximately 15 μm below the freezing surface and become progressively displaced deeper in the tissue. × 4,400.

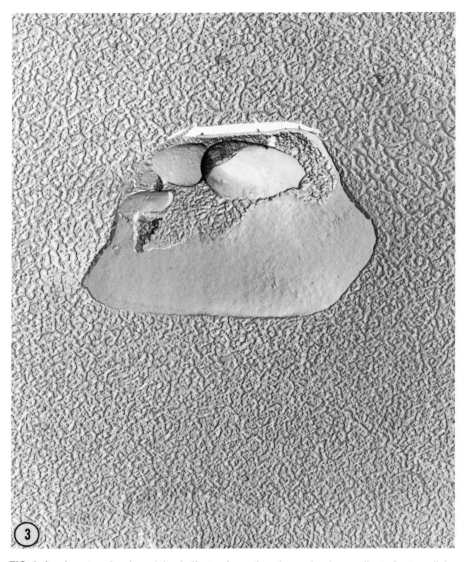

FIG. 3. Amebocytes showing minimal effects of poor freezing and a size gradient of extracellular ice crystals. Based on the size of these crystals, this cell is 15–20 μm below the freezing surface, but the fracture plane is tilted with respect to the freezing plane as indicated by the larger size of the ice crystals at the top of the picture (the steepness of the ice crystal gradient can vary slightly, even within the same block). A subtle effect of poor freezing is the roughness on the plasmalemmal E face of this cell, more pronounced on the right than the left. × 25,500.

increased with distance from the freezing surface (Figs. 3–8). However, these deformations in membranes could not be attributed to extracellular ice crystals impinging on them because individual crystals were larger than the deformations. Instead, the size of the deformations was closely matched to the size of the intracellular ice crystals, which in places appeared to impinge on the membrane beneath the deformations (Fig. 8).

The progressive increase in the size of ice crystals deeper in the tissue was only applicable to the extracellular spaces; intracellular ice crystals were always smaller than extracellular ones. However, gradients of size were present within cells (Fig. 3). Each amebocyte contained a gradient of ice crystals similarly oriented away from the freezing block. Because it is unlikely that cells were somehow cooled faster than the extracellular fluid, the tendency for ice crystals to be smaller inside cells must be related to properties of the cell interior which would affect ice crystal initiation or growth rates. Curiously, the membranes of intracytoplasmic secretory granules exposed to these same ice crystals were not deformed (Fig. 8), so different membrane properties, or contents of intracellular organelles may be factors in ice crystal damage (4).

Our experience in freezing other tissues has supported the general principle that intracellular ice crystals are smaller than extracellular ones. Furthermore, there is a tendency for this difference to be greater in smaller cells or cellular processes than in larger ones where ice crystals may be difficult to avoid. For instance, the glial processes around the squid giant axon contain very small ice crystals whereas ice crystals are much larger in the immediately adjacent giant axon (6). Similarly, muscle cells from frogs typically contain much larger ice crystals than the adjacent nerve terminals and Schwann cell processes (7). Plausible explanations for an inverse relationship between the size of intracellular processes and the size of the ice crystals found inside could depend on restraint of movements during crystal growth or on small amounts of evaporative drying (12). However, a moist chamber is used to minimize drying while the specimen is mounted on the freezing machine (7).

Other changes occurred in the plasma membranes of amebocytes deformed during ice crystal formation. Intramembrane particles appeared to be less frequent on the apices of deformations on P faces of amebocytes (Fig. 7). The E faces also were differentiated by many elongated "particles" and round-textured regions associated with the membrane deformations. These changes in particle structure may indicate an increased tendency for plastic deformation of intramembrane components as the sizes of the extra-membrane ice crystals increase (14).

Deeper than 50 μm in the tissue, where ice crystal damage was very extensive, the fracture plane tended to skip in and out of the plasma membrane, leaving membrane fragments or holes in the fracture face (Fig. 8). Also, linear deformations appeared on P faces which could be mistaken for ridges associated with tight junctions. Many of the changes discussed here have been seen in other

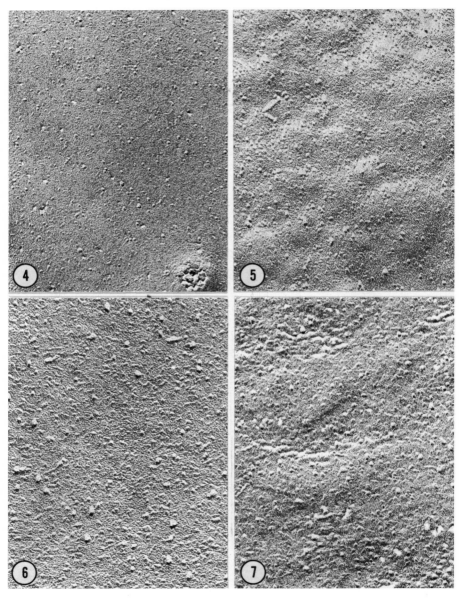

FIGS. 4–7. Fracture faces (P at top, E below) showing examples of best freezing, less than 10 μm from the surface (left column) and poorer freezing 25–50 μm from the surface (right column). Differential features of good freezing are that the well frozen fracture faces are smooth whereas they have a cobblestone appearance in poorly frozen material. Intramembrane particles are more pleomorphic in the poorly frozen examples and tend to be more sparse or less prominent, on the tops of membrane convexities. × 93,500.

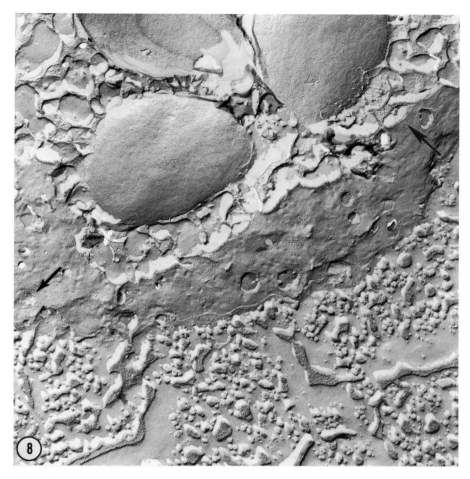

FIG. 8. Portion of a frozen cell more than 50 μm from the freezing surface. In addition to a cobblestone deformation of the plasmalemma, which appears to correspond to impinging *intracellular* ice crystals (large arrow), the P-face of this cell now has numerous holes and places where the plane of fracture left the cell membrane (small arrow, left). The membranes of secretory granules (above) are, however, spared the effects of the intracellular ice crystals. × 42,500.

tissues (5,8). It should be emphasized that the extent to which they are directly related to ice crystal damage is uncertain.

With the amebocyte preparation, cells must be selected from the first 10–15 μm below the freezing surface to be free of the artifacts listed above. These depths correspond to extracellular ice crystals approximately 20–30 nm in diameter, a dimension almost too small to distinguish clearly from normal hemocyanin distributions. Extrapolating from the data with cubes of liver shown in Fig. 1, amebocytes at these depths should be frozen in a few milliseconds after contact with the cold copper block. Of course, more sensitive statistical tests (2) must be applied to determine whether subtle changes in particle distribution due to ice crystal growth could occur on this time scale. It is still not certain whether ice crystal growth or other factors associated with freezing have any effects on particle distribution or structure, even when freezing is "good."

Various factors affect the depth in rapid-frozen tissue which is free of ice crystals and other effects of freezing. The choice of a cooling method is important; the colder and more heat diffusive the coolant, the deeper the zone free of ice crystals (7,10). In practical terms, a liquid coolant, such as Freon 23, instead of a cold metal block may be satisfactory to freeze monolayers of cells a few μm thick (3), or droplets of cell suspensions (1). Coolants such as LN_2 which boil at low temperatures should obviously be avoided. Liquid coolants are applicable to cylindrical or spherical specimens, where the ice crystal gradient will have a different geometry. With a sufficiently small cylinder the freezing may actually improve in the center (15).

Factors which affect intracellular ice crystal formation will be of paramount importance. As mentioned above, the size of cells or cytoplasmic processes has a significant but unexplained effect on the size of intracellular ice crystals. Cytoplasmic water available for ice crystal growth could also vary in different types of cells; in general, the results with marine animals have been better than with comparable cells from mammals or frogs. The reported improvement in freezing with osmotic agents confined to extracellular spaces, or after slight evaporative drying may also be understandable in terms of decreasing cytoplasmic water (13). Finally, fixing tissue in aldehyde before freezing may make membranes less sensitive to certain artifacts, such as plastic deformation of intramembrane particles (14). The appearance of freezing artifacts in tissues treated with glycerin or other cryoprotective agents is beyond the scope of our experience, though many of the above considerations are likely to be applicable (14).

As direct freezing techniques become more widely used, criteria for recognizing the subtle artifacts of freezing damage will become crucial. For instance, membrane deformations and particle displacements, as illustrated in Fig. 5, might be taken for real structural changes in membranes; for example, in preparation for exocytosis. In fact, when we limited our study of exocytosis to amebocytes within 10–15 μm of the initial plane of freezing, no changes in particle distribution are found before the formation of an exocytotic stomata. Whether this is a general property of exocytosis in other cells remains to be seen, but the final conclusions regarding the initiation of exocytosis, as well as all other aspects of membrane structure, must *at least* be based on cells free of the obvious artifacts of freezing outlined in this paper.

REFERENCES

1. Bachmann, L., and Schmitt, W. W. 1971. Improved cryofixation applicable to freeze etching. *Proc. Natl. Acad. Sci. U.S.A.*, 68:2149–2152.

2. Cohen, S., and Pumplin, D. 1979. Clusters of intramembrane particles associated with binding sites for α-bungarotoxin in cultured chick myotubes. *J. Cell Biol., (in press).*
3. Costello, M. J., and Corless, J. M. 1978. The direct measurement of temperature changes within freeze-fracture specimens during rapid quenching in liquid coolants. *J. Microsc.,* 112:17–37.
4. Dempsey, G. P., and Bullivant, S. 1976. A copper block method for freezing non-cryoprotected tissue to produce ice-crystal-free regions for electron microscopy. I. Evaluation using freeze substitution. *J. Microsc.,* 106:251–260.
5. Dempsey, G. P., and Bullivant, S. 1976. A copper block method for freezing non-cryoprotected tissue to produce ice-crystal-free regions for electron microscopy. II. Evaluation using freeze-fracturing with a cryo-ultramicrotome. *J. Microsc.,* 106:261–272.
6. Henkart, M. P., Reese, T. S., and Brindley, F. J. 1978. Endoplasmic reticulum sequesters calcium in the squid giant axon. *Science,* 202:1300–1302.
7. Heuser, J. E., Reese, T. S., Dennis, M. J., Jan, Y., Jan, L., and Evans, L. 1979. Synaptic vesicle exocytosis captured by quick freezing and correlated with quantal transmitter release. *J. Cell Biol.,* 81:275–300.
8. Heuser, J. E., Reese, T. S., and Landis, D. M. D. 1976. Preservation of synaptic structure by rapid freezing. *Cold Spring Harbor Symp. Quant. Biol.,* XL:17–24.
9. Meryman, H. T. (Ed.). 1966. Cryobiology. Academic Press, New York.
10. Nei, T. 1976. Review of the freezing techniques and their theories. In: *Recent Progress in Electron Microscopy of Cells and Tissues,* edited by E. Yamada et al., pp. 213–243. University Park Press, Baltimore.
11. Ornberg, R. L., and Reese, T. S. 1978. Early stages of exocytosis captured by rapid-freezing. *Neurosci. Abstracts,* 4:247.
12. Rehbum, L. I., and Sander, G. 1971. I. Conditions for avoidance of intracellular ice crystallization. *Am. J. Anat.,* 130:1.
13. Sleytr, U. B., and Robards, A. W. 1977. Plastic deformation during freeze-cleavage: a review. *J. Microsc.,* 110:1–25.
14. Staehelin, L. A., and Bertaud, W. S. 1971. Temperature and contamination dependent freeze-etch images of frozen water and glycerol solutions. *J. Ultrastruct. Res.,* 37:146–168.
15. Van Venrooij, G. E. P. M., Aertsen, A. M. H. J., Hax, W. M. A., Ververgaert, P. H. J. T., Verhoeven, J. J., and Van Der Vorst, H. A. 1975. Freeze-etching: Freezing velocity and crystal size at different locations in samples. *Cryobiology,* 12:46–61.

Freeze Fracture: Methods, Artifacts, and Interpretations, edited by J. E. Rash and C. S. Hudson. Raven Press, New York © 1979.

Controlled Contamination of Freeze-Fractured Specimens

Russell L. Steere, Eric F. Erbe, and J. Michael Moseley

Plant Virology Laboratory, Plant Protection Institute, AR, SEA U.S. Department of Agriculture, Beltsville, Maryland 20705

In his original paper on freeze etching, Steere (8) was concerned with potential artifacts that might be encountered in the freeze-etch technique. He considered the technique useful in spite of such possible artifacts which would, most likely, be quite different from those produced by thin-section techniques. The original approach required the use of an etching step due to condensation of moisture onto the fractured surface as the specimen was introduced into the vacuum unit. The introduction of freeze fracturing as contrasted with freeze etching (1), the development of a new freeze-etch unit (9) and of the double shroud (10), and the introduction of complementary tooling (12,13) have improved our ability to study artifacts associated with these techniques. Freeze fracturing of complementary specimens within the shroud of our freeze-etch unit exposes the two apposed fracture faces to potential surface contaminants. Fortunately, the shroud, in this instance, is a sufficiently good cold trap that when the specimen temperature is maintained at $-196°C$ in a vacuum of 10^{-6} torr and is properly oriented within the shroud, no detectable difference is seen in fracture faces of specimens shadowed immediately and those shadowed 5 min later. With this system, one can devise methods of introducing known contaminants under controlled conditions and visualizing their effect on specimen detail.

Gross et al. (4) have described the faults of "conventional freeze etching" in systems in which unshrouded specimens are exposed to vacuums of 10^{-6} torr. Because there are fewer uncontrolled contaminating molecules in the entire vacuum chamber under ultrahigh vacuum conditions than in our shrouded system at 10^{-6} torr, their system does provide potentially more reliable conditions for the examination of controlled contamination of fractured specimens with the introduction of known contaminants (3). Nevertheless, they have ignored the value of liquid nitrogen-cooled shrouds and have gone to great expense to develop an ultrahigh vacuum system to provide anticontamination conditions at the specimen surface equivalent to those provided by our liquid nitrogen-cooled shroud. Their system, with its lengthy pump-down time and the need for a bake-out period, makes it much less versatile than a system that uses a cold shroud as an anticontamination device.

Since the major contaminants of freeze-fracture and freeze-etch specimens are H_2O, Freon (if used for freezing), and hydrocarbons from the vacuum system, we believe that our system provides adequate conditions for observation of controlled specimen contamination. Furthermore, the conditions examined here are those of an operating, versatile, freeze-fracture, freeze-etch module that is commercially available and not those of a special one-of-a-kind device.

MATERIALS AND METHODS

Specimens consisted of suspended dry or cultured yeast cells *(Saccharomyces cerevisiae)* or leaf tissue of tobacco (*Nicotiana tabacum* L. cv. Samsun) suspended in 25% glycerol plus 25% sucrose as described elsewhere (15).

All specimens were prepared in a modified Denton DFE-2 freeze-etch module[1] mounted on a Denton DV 503 vacuum system. Modifications to the DFE-2 freeze-etch module consisted of modified resistance evaporators, complementary tooling, double shroud with fracture bar behind window opening, carbon source lifted to same level as the platinum source, and alignment stops. (These modifications are incorporated in the current commercial models of the Denton DFE-3 freeze-etch module.)

Platinum shadows were deposited with the aid of a monitor (14) and carbon thickness determined by visual means. Biological remains were removed from the replicas with chromic–sulfuric acid cleaning solution (Fisher SO-C-88) (11). Cleaned, washed specimens were mounted on electron microscope grids with or without support films. They were examined in a JEOL JEM 100-B transmission electron microscope equipped with a 60° top-entry goniometer stage. Stereoelectron micrographs were obtained with 10° tilt between pictures of each pair. Since original electron micrographs have black shadows, they are considered positives in this laboratory. Contact negatives of the originals were made on Kodak medium contrast projector slide plates. These negatives were then used to make final prints with black shadows as observed in nature and as presented by Steere (8) in the original freeze-etch publication.

For the study of surface contaminants, several methods were applied:

(a) Specimens become contaminated by prefracturing during the freezing and transfer steps. We have observed that some specimens, upon freezing, develop fissures that extend well into the frozen block but not all the way through it. These fissures permit the contamination of the prefractured faces. We have examined the contaminants deposited on such faces of specimens frozen in Freon-22 (Genetron 22-monochlorodifluoromethane).

(b) Complementary specimens were fractured at − 196°C in such a manner that the apposed specimen halves were left in contact. The specimen stage was

---

[1] Mention of a trademark, proprietary product, or vendor does not constitute a guarantee or warranty of the product by the U.S. Department of Agriculture and does not imply its approval to the exclusion of other products of vendors that may also be suitable.

then warmed to $-105°C$ and held at that temperature for 5 min. This procedure was employed to permit potential contamination of one-half of a fractured specimen during sublimation or etching of its complement.

(c) Specimens fractured and held at $-196°C$ were exposed to contamination by ultrahigh-purity dry nitrogen gas, bone-dry CO_2 gas, water vapor, or laboratory air, each introduced into the system through the chamber vent. For these experiments, the fractured specimens were exposed by aligning them with the chamber vent through the access port of the shroud. As further controls of the system, fractured specimens were exposed through the access port for 5 min to a vacuum in which both the trap above the diffusion pump and the shroud were cooled to $-196°C$, the trap above the diffusion pump was cooled but the shroud was left at room temperature, or both were left at room temperature.

(d) A second specimen tube was installed in the front port beneath the carbon source so that its specimen stage was directly beneath the existing specimen stage. The carbon aperture of the shroud was removed to allow free passage of contaminants from the lower specimen tube to the fractured specimen surface. The distance between the source of contamination and the specimen surface was 62 mm. Water or Freon-22 was frozen on the stage of the second specimen tube and held at $-196°C$ until the specimen was ready for fracturing. The shroud and specimen tube within it were cooled to $-196°C$ and, after an adequate vacuum (10^{-6} torr) was obtained, the lower specimen tube was warmed to the desired temperature for sublimation of the specific contaminant. The specimen was then fractured and aligned through the carbon aperture tube with an unobstructed view of the contaminating source. Fractured specimens at -196, -120, or $-110°C$ were exposed for various time intervals to the specific contaminants.

RESULTS AND DISCUSSION

Yeast Preparations Partially Prefractured During Freezing in Freon and Fractured at $-196°C$

Yeast preparations immersed in Freon-22 at approximately $-160°C$ often develop fissures upon freezing that permit the exposure of prefractured faces to constituents of the Freon. These fissures can be visualized with a dissecting microscope before the specimens are introduced into the complementary specimen cap. They do not traverse the entire specimen; therefore, the specimen holders do not open prematurely. Upon freeze fracturing such specimens at $-196°C$, two distinct zones are observed: the prefractured zone containing cubic crystals on both fracture faces with tubular structures [the craters of Gross et al. (4)] (Fig. 1) on the P-faces, and those zones actually fractured at liquid nitrogen temperature that contain no such crystals but reveal the tubular structures with many scattered strands [presumed by some to be caused by

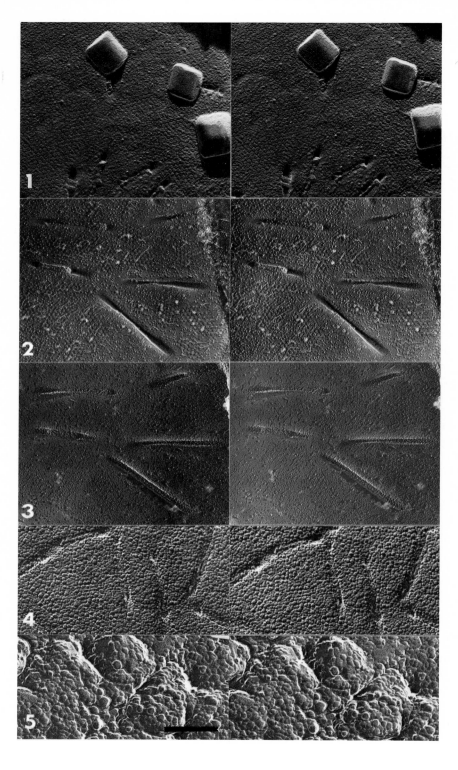

"plastic deformation" as reviewed by Sleytr and Robards (6)] on the P-face (Fig. 2) and ring-like depressions on the complementary E face (Fig. 3). The exact composition of the cubic crystals has not been determined, but they do sublime away freely at − 98°C and are not rapidly removed under conditions where the major constituents of Freon evaporate or sublime (− 105°C or colder). The crystals could be components of the Freon-22 itself or merely cubic ice crystals present in the liquid Freon in which the specimens were frozen.

Yeast Samples Cross Fractured But Not Opened Until Shadowing

Specimens from the firmly attached half of the specimen holders warmed to and held at − 105°C for 5 mins after being fractured at − 196°C reveal a cleanly etched surface in which the very fine ultrastructure of the lipid bodies (5) is clearly defined (Fig. 6). The complementary loose halves reveal a confluent contaminant except over the lipid bodies (Fig. 7). Apparently the loose halves of such specimen holders are in poor contact with the warmed halves, and sufficiently well protected from outside radiant heat by the cold shroud that they remain considerably colder than the firmly attached halves and serve as a cold trap for condensation of whatever sublimes from the warmer complement.

Yeast Cells Fractured at − 196°C, then Exposed to Introduced Contaminants

Control: Immediate replication. Controls consisted of yeast cells freeze fractured at − 196°C within liquid nitrogen-cooled shroud, shadowed, and replicated immediately. These revealed tubular particles [the crater-like particles of Gross et al. (4)] plus many fine strands on the P-face and ring-like depressions on the complementary E-face (Figs. 2 and 3).

Lack of contamination, with shroud and diffusion pump trap cold. Specimens fractured at − 196°C, etched for 5 min at − 95°C, recooled to − 196°C, and exposed to standard vacuum conditions (1 × 10^{-6} torr) through the specimen access port with the diffusion pump trap and shroud at − 196°, revealed no observable contamination (not shown).

Contamination, with diffusion pump trap cold but shroud at room temperature. Specimens fractured at − 196°C, etched for 5 min at 95°C, recooled to − 196°C, and exposed to the vacuum system through the specimen access port with the

---

FIGS. 1–5. Freeze-fractured yeast preparations. **1:** P-face contaminated with cubic crystals as a result of prefracturing during freezing in Freon-22. Tubular structures in paracrystalline arrays [the craters of Gross et al. (4) are clearly visible. **2:** P-face of yeast fractured at − 196°C, shadowed, and replicated immediately. Fibers are always present in addition to the tubules when specimens are fractured at − 196°C. **3:** E-face of the precise complement to Fig. 2 printed so that complementary parts are at the same relative positions in both prints. Ring-like depressions are complementary to the tubes of Fig. 2. **4:** Contamination resulting from exposure of the fractured specimen to vacuum within chamber with shroud and cold trap of diffusion pump at room temperature. **5:** Fractured yeast cell (− 196°C) contaminated by injection of 1 ml of H_2O into vacuum chamber through chamber vent. Figs. 1–5: Bar, 200 nm; ×71,000.

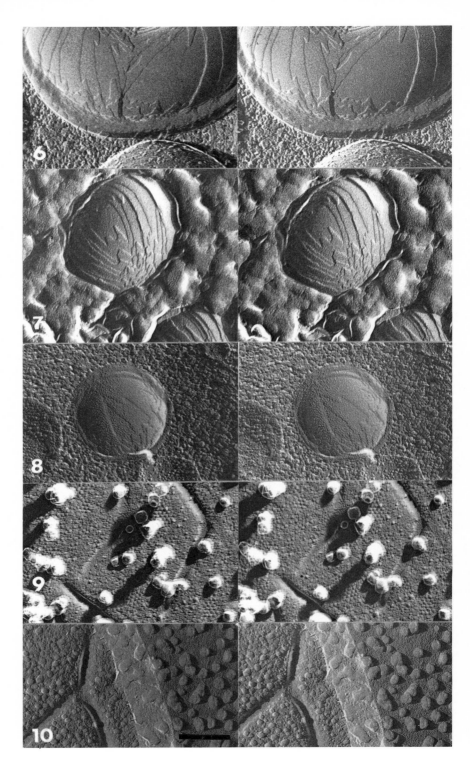

diffusion pump cold trap at $-196°C$ and shroud at room temperature revealed the presence of a uniform granular contaminant on the lipid bodies (Fig. 8) that was not observed in Figs. 6 and 7.

Contamination with shroud and diffusion pump trap at room temperature. Specimens fractured at $-196°C$, etched for 5 min at $-95°C$, then recooled to $-196°C$, and exposed to the vacuum system through the entry port for 5 min with diffusion pump trap and shroud at room temperature revealed many small globular deposits which obscure all fine structural details (Fig. 4).

Lack of contamination from dry nitrogen gas. Fractured specimens held at $-196°C$ and exposed to ultrahigh-purity dry nitrogen gas through specimen access port to a pressure greater than 50 millitorr with shroud and pump trap at $-196°C$ revealed no observable contamination. (not shown.)

Lack of contamination from dry CO_2. Fractured specimens held at $-196°C$ and exposed to bone-dry CO_2 gas through specimen access port revealed no observable contaminant when the chamber was repumped to 10^{-6} torr with shroud and pump trap at $-196°C$. Presumably, the "CO_2 crystals" previously described (10) were composed of contaminating H_2O deposited in the presence of CO_2. (not shown.)

Contamination from injected water vapor. Fractured specimens held at $-196°C$ and exposed through specimen access port to contamination by injection of 1 ml of H_2O into the chamber vent (shroud and pump trap at $-196°C$) revealed extensive deposits with the typical H_2O contaminant that obliterated all structural detail of the specimen (Fig. 5).

Contamination at $-196°C$ from injected room air. Fractured specimens held at $-196°C$ and exposed through specimen access port to a 3-sec burst of laboratory air through the ¼-inch toggle valve (shroud and pump trap at $-196°C$) were contaminated with aggregates of small crystals on an otherwise very lightly contaminated surface (Fig. 9).

Contamination at $-120°C$ from injected room air. Specimens fractured at $-196°C$, warmed to and held at $-120°C$ during exposure through specimen access port to a 3-sec burst of laboratory air through the ¼-inch toggle valve, then recooled to $-196°C$ for shadowing and replication (shroud and pump cold trap at $-196°C$) were contaminated with many globular deposits which obscure the fine structural details of membrane surfaces (Fig. 10). Similar but larger globular or quasi-crystalline deposits covered the nonmembrane areas.

FIGS. 6–10. Freeze-fractured yeast preparations. **6:** Fine structure of lipid bodies on firmly attached side of specimen holder etched for 5 min at $-105°C$ with loose half of specimen remaining in contact until time of shadowing. **7:** Area from fracture surface of specimen half complementary to Fig. 4 with contamination everywhere except on lipid bodies. **8:** Lipid body of specimen held at $-196°C$ and contaminated by exposure to vacuum within chamber with shroud at room temperature and cold trap of diffusion pump at $-196°C$. **9:** P-face of yeast cell fractured and held at $-196°C$ during contamination by introduction of a 3-sec burst of laboratory air. **10:** Fractured yeast cell held at $-120°C$ during contamination by introduction of a 3-sec burst of laboratory air. Figs. 6–10: Bar, 200 nm; ×71,000.

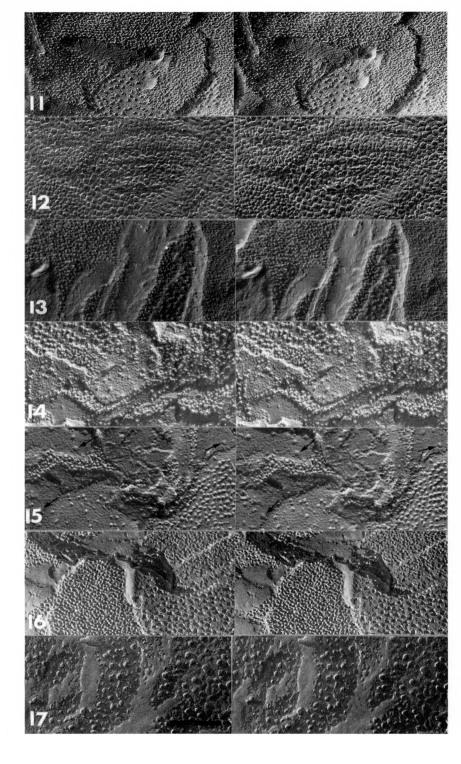

Controlled Sublimation of Contaminants from Second Cold Stage and Consequent Contamination of Exposed Fractured Surfaces (Chloroplasts of Unfixed Tobacco Tissue as Specimens)

Figures 11–17 were freeze-fractured chloroplasts within unfixed tobacco leaf tissue cryoprotected with 25% glycerol–25% sucrose in 0.2 M phosphate buffer, pH 7.0

Control preparations of tobacco chloroplasts. As controls for this series, small samples of tobacco tissue were frozen in Freon-22 and freeze fractured at − 196°C, then shadowed and replicated immediately with shroud and pump cold trap at − 196°C. Other samples were frozen by immersion in liquid nitrogen. Since the original samples were suspended in a 25% glycerol–25% sucrose solution in buffer, no sign of crystals developed in the samples frozen more slowly in the liquid nitrogen and both control specimens had the same general appearance (Fig. 11).

Contamination at − 196° C from subliming water vapor plus Freon-22. Specimens for this test were frozen in Freon-22 and placed in the complementary specimen cap, then transferred to the cold stage at − 150°C. The lower cold stage was held at − 196°C and distilled H_2O was frozen onto it. As Freon-22 evaporated from specimens held at − 120 to − 150°C during pumpdown, it condensed out on the lower ice-coated cold stage. When a good vacuum was obtained, specimen temperature was lowered to and held at − 196°C. The lower stage was warmed to − 95°C and specimens were fractured and exposed through the carbon tube of the shroud for 1 min to contamination by subliming vapors from the lower stage (shroud and pump cold trap at − 196°C). The fracture faces of these specimens were completely covered with globular-to-crystalline deposits which obliterated all but the gross structural details (Fig. 12).

Contamination at − 196° C from subliming water vapor without Freon-22 (1 min.). When the plant tissue samples are frozen in liquid nitrogen rather than in Freon so that no Freon is present in the vacuum system (other conditions being the same as for the preceding experiment) contamination consists of only slight reduction in the resolution of the smallest depressions and slight enlargement of membrane-associated particles (Fig. 13). This is quite a contrast to

FIGS. 11–17. 11: Control: freeze fractured at − 196°C, shadowed, and replicated immediately after fracturing. 12: Contamination of specimen held at − 196°C during exposure for 1 min to sublimate from frozen H_2O onto which Freon-22 had condensed. 13: Contamination of specimen exposed at − 196°C under conditions like those of Fig. 12, except with no Freon present. 14: Contamination of specimen held at − 196°C during exposure for 5 min to subliming H_2O from second specimen stage at − 95°C, no Freon present. 15: Contamination of specimen held at − 120°C for 1 min during exposure to H_2O and Freon sublimation (− 95°C) from second specimen stage. Contaminants *(arrows)* similar to those of Fig. 12 appear only on nonmembrane areas. 16: Contamination of specimen held at − 110°C during 3-min exposure to subliming H_2O from second specimen stage, no Freon present. 17: Contamination of specimen prepared simultaneously with that used for Fig. 16 (no Freon present). Figs. 11–17: Bar, 200 nm; ×71,000.

the severe contamination observed when the system contains Freon. Therefore, we suspect that the contaminant shown in Fig. 12 is condensed Freon. It is possible, however, that this contaminant is a mixture of Freon and H_2O or only H_2O condensed out on the specimen surface in the presence of Freon. Additional experiments are needed to establish the precise interaction of Freon and H_2O on cold surfaces.

Contamination at $-196°C$ *from subliming H_2O vapor (5 min).* Samples frozen in liquid nitrogen, fractured, and held at $-196°C$ during 5-min exposure to contamination through carbon tube of the shroud by subliming H_2O appeared rather normal at first. At higher magnification, however (Fig. 14) it becomes clear that all fine structural detail is obscured and the membrane-associated particles are enlarged. These images resemble the "cobblestones" or "orange peel" surfaces described by Staehelin and Bertrand (7) and Deamer et al. (2).

Contamination at $-120°C$ *by subliming H_2O plus Freon-22.* Samples frozen in Freon as above, but with specimen temperature raised to $-120°C$ after fracturing, and exposed for 1 min through the carbon tube of the shroud to the subliming water and Freon (shroud and pump cold trap at $-196°C$), then recooled to $-196°C$ for shadowing and replication (Fig. 15), revealed contaminants similar to those in Fig. 12 (the other Freon-frozen specimens). In this instance, however, fine detail is not obscured. The contaminating bumps are slightly larger than membrane-associated particles but are found only on non-membrane surfaces *(arrows).* Apparently the specimen temperature at $-120°C$ is borderline for condensation of Freon onto the etching surface which might be slightly colder than the nonetching membrane faces.

Contamination at $-110°C$ *by subliming H_2O: Local differences.* Samples frozen in liquid nitrogen as in (c), but with specimen temperature raised to $-110°C$ after fracturing for the 3-min exposure to H_2O subliming from the lower cold stage at $-95°C$, and subsequently recooled to $-196°C$ for shadowing and replication reveal fracture faces in which the background depressions are obliterated and the membrane-associated particles are greatly enlarged into angular particles (Fig. 16).

Another specimen prepared simultaneously with that of Fig. 16 appears to be much more grossly contaminated with large quasi-crystalline particles scattered over the membrane faces (Fig. 17). Perhaps a slight difference in angle of the fracture face or alignment of specimen through the shroud opening with the contaminating source is responsible for such a drastic difference in the appearance of these two specimens.

SUMMARY

The experiments described and illustrated here served to convince us that the double-shroud system is even a better cold trap than we had anticipated. It was found necessary to remove the carbon aperture from the shroud, to expose the fractured specimen through the specimen access port of the outer chamber, keep the shroud at room temperature, or expose a colder half of the fractured specimen to its subliming (etching) complement to provide reasonable rates of contamination. Of particular importance, we believe, is the differences in observed contamination dependent on the temperature of the specimen being contaminated; what we believe to be Freon contamination is very pronounced on specimens held at $-196°C$ and very minimal on specimens held at $-120°C$. We have also seen that different specimens exposed simultaneously under identical conditions can be contaminated to different degrees. This suggests that slight differences in the angle of exposure of specimen surfaces to the contaminating source may affect the degree of contamination.

REFERENCES

1. Bullivant, S., and Ames, A., 3rd. (1966): A simple freeze-fracture replication method for electron microscopy. *J. Cell Biol.,* 29:435–447.
2. Deamer, D. W., Leonard, R., Tardieu, A., and Branton, D. (1970): Lamellar and hexagonal phases visualized by freeze-etching. *Biochem. Biophys. Acta,* 219:47–60.
3. Gross, H., and Moor, H. (1978): Decoration of specific sites on freeze-fractured membranes. *Proc. 9th Int. Congr. Electron Microscopy, Toronto,* edited by J. M. Sturgess. Vol. II, pp. 140–141. Microscopical Society of Canada, Toronto, Ontario, Canada.
4. Gross, H., Bas, E., and Moor, H. (1978): Freeze-fracturing in ultrahigh vacuum at $-196°C$. *J. Cell Biol.,* 76:712–728.
5. Moor, H., and Mühlethaler, K. (1963): Fine structure in frozen-etched yeast cells. *J. Cell Biol.,* 17:609–628.
6. Sleytr, U. B., and Robards, A. W. (1977): Plastic deformation during freeze-cleavage: A review. *J. Microscopy,* 110:1–25.
7. Staehelin, L. A., and Bertrand, W. S. (1971): Temperature and contamination dependent freeze-etch images of frozen water and glycerol solutions. *J. Ultrastr. Res.* 37:146–168.
8. Steere, R. L. (1957): Electron microscopy of structural detail in frozen biological specimens. *J. Biophys. Biochem. Cytol.,* 3:45–60.
9. Steere, R. L. (1969a): Freeze-etching simplified. *Cryobiology,* 5:306–323.
10. Steere, R. L. (1969b): Freeze-etching and direct observation of freezing damage. *Cryobiology,* 6:137–150.
11. Steere, R. L. (1973): Preparation of high-resolution freeze-etch, freeze-fracture, frozen-surface, and freeze-dried replicas in a single freeze-etch module, and the use of stereo electron microscopy to obtain maximum information from them. In: *Freeze-Etching Techniques and Applications* Chapter 18. edited by E. L. Benedetti and P. Favard, pp. 223–255. Société Française de Microscopie Electronique, Paris.
12. Steere, R. L., and Moseley, M. (1969): New dimensions in freeze-etching. *Proc. 27th Ann. Mtg., Electron Microscopy Soc. Am.,* edited by C. J. Arceneaux, pp. 202–203. Claitor's Publ. Div., Baton Rouge.
13. Steere, R. L., and Moseley, M. (1970): Modified freeze-etch equipment permits simultaneous preparation of 2–10 double replicas. *Proc. 7th Int. Congr. Electron Microscopy, Grenoble.* edited by P. Favard, pp. 451–452. Société Française de Microscopie Electronique, Paris.
14. Steere, R. L., Erbe, E. F., and Moseley, J. M. (1977): A resistance monitor with power cut-off for automatic regulation of shadow and support film thickness in freeze-etching and related techniques. *J. Microsc.,* 111(Pt. 3):313–328. Société Française de Microscopie Electronique, Paris.
15. Steere, R. L., Moseley, J. M., and Erbe, E. F. (1975): Use of sucrose–glycerol density gradients in the preparation of living cells for high-resolution complementary freeze-fractured specimens. *Proc. 33rd Ann. Mtg., Electron Microscopy Soc. Am.,* edited by G. W. Bailey, pp. 616–617. Claitor's Publ. Div., Baton Rouge.

Freeze Fracture: Methods, Artifacts, and Interpretations, edited by J. E. Rash and C. S. Hudson. Raven Press, New York © 1979.

Sources and Rates of Contamination in a Conventional Balzers Freeze-Etch Device

John E. Rash, William F. Graham, and C. Sue Hudson

Department of Pharmacology and Experimental Therapeutics, University of Maryland School of Medicine, Baltimore, Maryland 21201

One of the most perplexing problems confronting every investigator using freeze-fracture methodologies is the apparent capriciousness involved in consistently obtaining replicas that are devoid of detectable surface contamination. "Contamination," defined as any material deposited on freshly cleaved surfaces prior to replication, has been shown to consist primarily of water vapor, either in the form of frozen microdroplets ("crystallites") or as a featureless blanket that obliterates all molecular detail (7,10). Additional substances present in the bell jar that contribute minimally to contamination at conventional specimen cleaving temperatures include glycerol (dissolved in the water subliming from specimen chips), Freon or isopentane (freezing mixtures), residual carbon dioxide, and, perhaps, the vapors arising from the hot oil or ion diffusion pumps (R. L. Steere et al., 1979, *this volume*). In addition, a detailed analysis of water vapor contamination at much lower cleaving temperature ($-196°C$) and at much lower bell jar pressure (10^{-9} torr) is described by H. Gross *(this volume).*

Detailed analyses of water vapor contamination of freeze-etched artificial lipid bilayers (7) and of glycerol–water mixtures cleaved and replicated at conventional temperatures and pressures (10) revealed several morphologies of the deposited crystallites and identified water vapor streaming from the chip-coated knife edge as the major source of contamination under conventional cleaving condition. However, neither group presented data concerning other intrinsic sources of water vapor contamination or attempted to describe the rate of alteration of particle size, number, or distribution in biological membranes. Nevertheless, these reports provided a rational basis for recognizing and minimizing surface contamination.

In this study, we attempt to (a) identify the most common major sources of contamination normally present within the Balzers freeze-fracture device, (b) determine the rate of surface contamination attributable to each source, (c) demonstrate the characteristic alterations of particle and surface morphology attributable to each, and (d) suggest simple methods for minimizing artifacts of contamination.

MATERIALS AND METHODS

The experiments described below were carried out in a standard Balzers 360 M freeze-etch device, evacuated by an oil diffusion pump backed by a mechanical rotary oil pump. The device was not equipped with a Meissner trap. Three intrinsic sources of water vapor contamination were measured using specimens cleaved at $-105°C$. These included: (a) residual and/or naturally subliming water vapor normally present at indicated bell jar pressures of 10^{-6} to 5×10^{-3} torr, (b) water vapor subliming from specimen chips on the knife edge positioned at distances of 2, 5, and 20 mm from the freshly cleaved specimen, and (c) water vapor deposited as "self-contamination" following premature partial fracture. The effect of specimen temperature on rate and morphology of water-vapor deposition at the conventional pressure of 10^{-6} torr was determined at specimen-cleaving temperatures of -105, -115, and $-150°C$.

Because the relative rates of contamination of hydrophilic versus hydrophobic surfaces are not yet established at ultralow temperatures, we have devised two semi-quantitative methods for measuring contamination thickness: we have measured the radius (half diameter) of artifactural "pseudoparticles" deposited on normally smooth portions of membranes, and we have estimated (with a micrometer eyepiece) the average increase in radius of certain easily recognized membrane particles (e.g., the "square arrays"). The measurements, of necessity, represent subjective evaluations of several variables and are not intended to represent precise values for all membranes or all particles. The rate of surface contamination (measured in $Å/sec$) was determined measuring contamination thickness at three points during the initial nearly linear phase of deposition (0.1, 5, and 10 sec) and plotting the slope of the curve. These relatively brief exposure times were selected for rate measurements because (a) significant contamination may occur during the conventional 5 to 10 sec required to examine the several freshly cleaved surfaces and to heat the electrode for platinum deposition, (b) contamination rates from some sources are so high that surface detail may be obliterated in only a few seconds, and (c) methods are now available for reducing exposure time to much less than 1 sec.

To measure the contamination occurring during these relatively brief periods of exposure, the electron-beam gun for platinum was preheated (1800 V, 70 μA) for 3.8 sec immediately preceding the final cleaving of the specimen. An electrically activated shutter (as shown in Fig. 1, M. H. Ellisman and L. A. Staehelin, *this volume*) prevented premature deposition of platinum on the specimen and any possible surface melting by radiant heat during the electrode warm-up period. The shutter was opened automatically 3.8 sec after activation of high voltage to the platinum gun. The final cleave was timed to coincide with the opening of the shutter ($\pm$ 100 msec.) To obtain reproducible exposures of 5 and 10 sec, the specimens were cleaved at 1.2 or 6.2 sec *before* the high voltage was activated. Shadow times were 1 to 5 sec.

To standardize the results, the bell jar was evacuated to 10^{-6} torr prior to each cleave; the main-plate valve was closed (to preclude back streaming of diffusion and/or rotary pump oils), and the bell jar pressure was allowed to rise slowly to the desired pressure. (Presumably, the rise in pressure is due primarily to sublimation of water from the specimens and from the hoar frost deposited on the specimen stage during specimen loading.) In all cases, the knife was chilled for at least two complete cooling cycles before cleaving was initiated; three to six shallow cleaves were completed rapidly. The final cleave was made during the final seconds of the last cooling cycle.

To characterize contamination of a relatively flat and smooth biological membrane, the E- and P-face images (2) of rat extensor digitorum longus (EDL) muscle plasmamembranes were examined. (For descriptions of particle morphology in rat muscle sarcolemmas, see refs. 8 and 9.) Replicas were photographed in a Siemens Elmiskop 101 operated at 80 kV. Particle diameters and heights were measured on $\times 100,000$ negatives using a $\times 10$ micrometer eyepiece.

RESULTS AND DISCUSSION

Use of Automatic Shutter for Simultaneous Cleaving and Shadowing

The value of the shutter in obtaining virtually contamination-free replicas of demonstrably high quality is readily apparent in replicas of the sarcoplasmic reticulum of mammalian skeletal muscle. The sarcoplasmic reticulum E- and P-faces are universally described as exhibiting complete noncomplementarity of membrane faces (1,5,6,9). When cleaved at conventional temperatures (-105 to $-115°C$) and replicated after the conventional delay of 5 to 10 sec (Fig. 1a), the sarcoplasmic reticulum P-face is described as containing numerous 80-Å particles (the "rough" face), whereas the E-face is described as being completely devoid of pits (the "smooth" face). However, when the interval between cleaving and replication is reduced to less than 1 sec (by using the shutter), the E-faces appear clearly and distinctly pitted (Figs. 1b and c), thereby revealing the near complementarity of these membrane E- and P-faces. Whether the obliteration of E-face pits results from extremely rapid surface contamination (i.e., from above) or by aperture occlusion by water vapor subliming through these holes in the membrane (i.e., from below) has yet to be determined. (The possibility of delayed plastic deformation of the lipid molecules forming the annulus of the pit seems unlikely, particularly since the pits also may be visualized after 30 sec of etching. Not shown.) Regardless of the precise source of contamination, the value of the shutter for minimizing rate-limited alterations of membrane infrastructure is clearly evident.

With the demonstrated ability to replicate consistently within 100 msec of the cleaving process, we initiated a systematic analysis of the deposition rates and distinctive morphologies of the various intrinsic sources of surface contamination, using the E- and P-face images of rat EDL nonjunctional sarcolemma as a test model. Within the E- and P-faces are characteristic "square arrays"

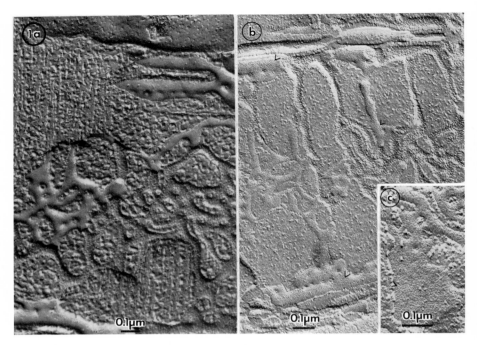

FIG. 1. Sarcoplasmic reticulum replicated **a:** about 10 sec after cleaving and **b** and **c:** 1 sec after cleaving. Compare the smooth E-faces in **a** with the distinctly pitted E-faces in **b**. Noncomplementarity of membrane faces in **a** is a reliable criterion for recognizing artifacts of the cleaving and replication process (see ref. 4). **a:** ×50,000; **b:** ×55,000; **c:** ×80,000.

of pits and particles, as well as other easily identified classes of particles that can be used to assess alterations resulting from water vapor deposition.

Contamination by Residual Water Vapor (10^{-6} to 5×10^{-3} torr)

The source most commonly assumed to contribute to surface contamination of specimens cleaved a -100 to $-150°C$ is the residual water vapor in the bell jar (cf., S. Böhler, *this volume;* H. Gross, *this volume*). While maintaining the specimen at $-105°C$, the effect of varying exposure times to bell jar pressures of 10^{-6} to 5×10^{-3} torr were investigated (Fig. 2). At 10^{-6} torr, no detectable differences were observed at 0.1-, 5-, or 10-sec exposure. Surprisingly, even at the extremely poor bell jar vacuum of 10^{-4} torr, only a very thin, nonuniform deposit of water vapor was detected, recognizable primarily by a moderate increase in the height and diameter of some particles (Fig. 2, upper right panel). Other particles and many E- and P-face pits (including the "square arrays") appear essentially unaltered. Moreover, the central portions of the distinct E-

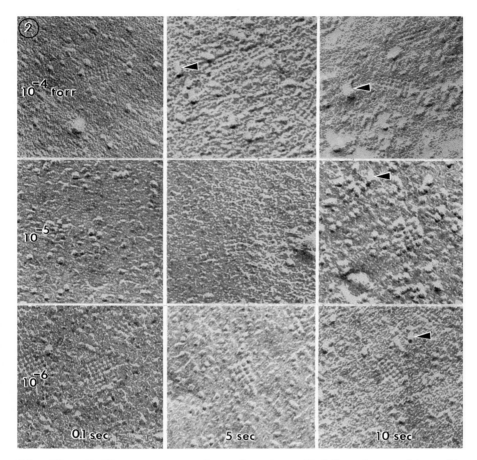

FIG. 2. Effect of bell jar pressure on specimens cleaved at − 105°C. Note absence of detectable contamination after 10-sec exposure at 10⁻⁶ torr. Particles were of increased diameter and pits were partially obliterated after 5- and 10-sec exposure at 10⁻⁴ torr. ×135,000.

face arrays of pits served as a "benchmark" against which the relative thickness of contamination deposited on the adjacent membrane face was compared. Because not all particles or pits were affected equally, estimates of the "average" increase in diameter of particles were plotted as a function of time (Fig. 3). These estimates of contamination rates over a wide range of pressures provide strong support for earlier suggestions that *the residual gases in the bell jar are not the primary source of severe surface contamination of specimens cleaved at − 105° C.* Moreover, these low rates of water vapor deposition were not entirely unexpected, since freeze drying is routinely accomplished under much poorer vacuum conditions (i.e., at 10 to 10⁻³ torr).

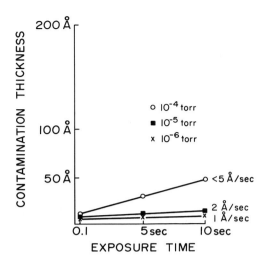

FIG. 3. Contamination rates at a specimen temperature of $-105°C$ attributable to various bell jar pressures. Excessive contamination rate observed at 10^{-4} torr. Rate measured in Å contamination per second of exposure.

Contamination at Lower Specimen Temperatures

Although very little specimen contamination was detected at a cleaving temperature of $-105°C$ and exposures for 10 sec at pressures less than 10^{-5} torr, at much colder specimen temperatures the freshly cleaved surface would be expected to serve as a "cold trap" to which water molecules would become frozen or adsorbed very quickly (cf., S. Böhler, Fig. 8, *this volume*). To assess the rate of contamination at other conventional cleaving temperatures, specimens were cleaved at -105, -115, and $-150°C$ and exposed for 0.1, 5, and 10 sec to a constant bell jar pressure of 10^{-6} torr (Fig. 4). No contamination was detected at $-105°C$. Even after 10 sec exposure at $-115°C$, particles and pits were distinct but particles appeared somewhat taller and of slightly increased diameter, perhaps 20 to 30 Å larger (Fig. 5). At $-150°C$, however, both E- and P-faces quickly developed a relatively uniform "cobblestone" appearance of 40 to 80 Å pseudoparticles (Fig. 5). The uniform granularity apparently reflects the rapid but nonspecific contamination of all surfaces by residual moisture present in the relatively homogeneous vacuum of the bell jar. Thus, we conclude that very low *specimen cleaving temperatures should be avoided unless elaborate precautions are taken to prevent contamination.* If ultralow specimen temperatures must be employed in order to minimize plastic deformation (for example, S. Kirchanski et al., *this volume*), additional protective methods must be instituted, including greatly reducing exposure time by the use of the shutter, protective cold traps and cooling shrouds that reduce local water-vapor pressure (cf., S. Bullivant, *this volume;* R. L. Steere et al., *this volume*), and/or *ultralow vacuums* (cf. H. Gross, *this volume*).

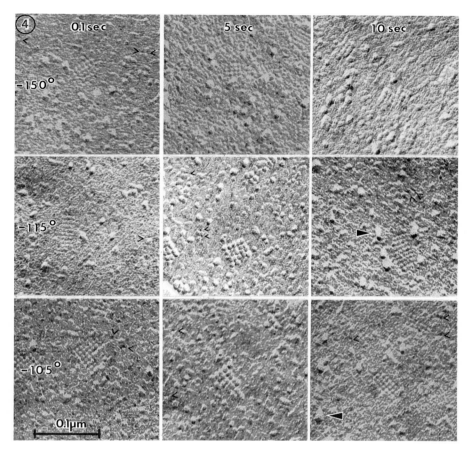

FIG. 4. Effect of specimen cleaving temperature on contamination rate. Constant bell jar pressure of 10^{-6} torr. (Knife edge 20 mm from specimen stage.) Slightly increased rate of contamination after 10-sec exposure at $-115°C$ reflected in increased particle height. Significantly increased contamination rate at $-150°C$ detected as evenly distributed fine-grain "cobblestone" or "orange-peel" effect. Pits obliterated, particles of increased diameter. ×130,000.

Ice Chips on the Knife Edge as a Major Source of Contamination

The preceding images were only slightly to moderately contaminated, yet we have all seen highly contaminated surfaces from specimens cleaved at $-105°C$ or higher. If residual water vapor deposited at conventional bell jar pressures and specimen temperatures is not the primary source, what is? Based on the careful studies of contamination of glycerol–water mixtures by Staehelin and Bertaud (10), we reexamined the contribution of local internal sources of water vapor, especially the ice chips on the knife edge (Fig. 6). At a pressure

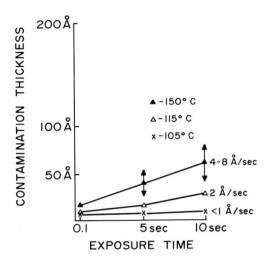

FIG. 5. Contamination rate for specimens cleaved at − 105, − 115, and − 150°C and exposed to a conventional bell jar pressure of 10⁻⁶ torr. Excessive rate of contamination at − 150°C would necessitate elaborate protective measures.

of 10⁻⁶ torr and a specimen temperature of − 105°C, several shallow cleaves were made and the knife subsequently positioned at 2, 5, and 20 mm from the specimen. Under these conditions, rapid surface contamination is observed when the knife edge is held near the specimen (Fig. 7) with contamination decreasing at about the inverse square of the distance of the specimen from the knife edge (Fig. 8). It should be noted that with the knife at 2 mm contamination probably occurs throughout the period of platinum deposition. (Codeposi-

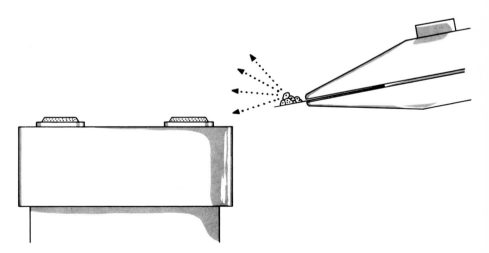

FIG. 6. Sublimation from ice chips adhering to knife edge. Poor thermal contact of specimen chips on knife edge combined with radiant energy within the room-temperature bell jar results in rapid warming of ice-glycerol "whiskers" and the violent release of water vapor and glycerol. Directionality of deposition reflected in uneven distribution of contaminants on curved surfaces.

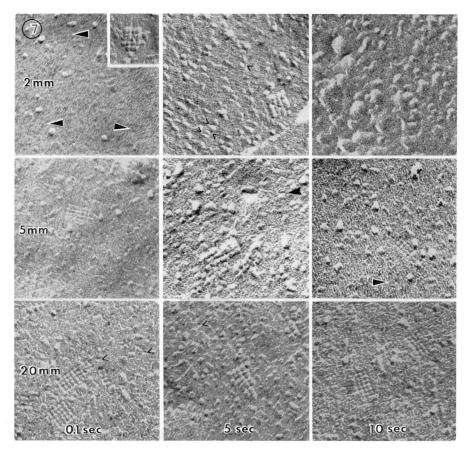

FIG. 7. Effect of specimen chips on knife edge positioned at 2, 5, and 20 mm from freshly cleaved specimen. (Constant pressure of 10^{-6} torr, constant specimen temperature of $-105°C$.) At 2 mm, contamination is extremely rapid. By 5 and 10 sec pits are obliterated and particles are enlarged. In some areas, very large diameter "pseudoparticles" obliterate all structural detail. The proximity of the knife edge coated with chips is confirmed as a primary source of surface contamination. ×130,000.

tion of contamination and platinum is reflected in Fig. 8). Interestingly, in these heavily contaminated replicas, the platinum shadows were especially delicate, often being partially removed or dislodged from the carbon support during replica cleaning. Apparently, the relatively thick layer of condensed water vapor prevents proper stabilization of the replica by the carbon support.

The very high rate of contamination in this experiment should be contrasted with the much lower values obtained following exposure to the residual water vapor in the bell jar at 10^{-6} to 10^{-4} torr. Apparently, poor thermal contact of the chips on the knife edge plus increased rates of sublimation of water

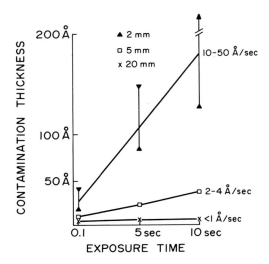

FIG. 8. Contamination rates attributable to ice chips on knife edge. Constant specimen temperature of − 105°C and bell jar pressure of 10^{-6} torr. With knife positioned immediately behind the specimens, contamination is extremely rapid. The knife edge *must* be kept at a maximum distance after *each* pass.

and glycerol from the numerous point sources (fracture edges) resulted in rapid but nonuniform surface contamination. (It should be remembered that the specimen chips are exposed to black body radiation from the bell jar which is at ambient room temperature.) Although similar contamination may result from deliberate warming of the knife edge (7), failure to properly position the knife after specimen cleaving is clearly the single most important and most likely source of contamination encountered by the otherwise careful investigator (10).

Self-Contamination

The final source of surface condensation investigated was "self-contamination" which occurs between closely opposed, incompletely fractured surfaces. In several instances, samples were observed to "rock" or "jiggle" as the knife passed. On the succeeding knife pass, the specimen cleaved very deeply and was shadowed immediately. These specimens invariably exhibited areas of gross contamination (Fig. 9). Since the precise instant of premature fracture was not known, we attempted to duplicate the effects of incomplete or premature partial fracture. Samples were partially fractured by lowering a fresh knife edge, initiating a partial break, and allowing the apposed specimen portions to self-contaminate for 10 sec. The fracturing and replication processes were then completed as usual. Prematurely cleaved membranes exhibited highly varied surface contamination, some laden with irregular patch-like pseudoparticles (Fig. 9a), some resembling "cobblestones" or "orange-peels" (Fig. 9b), and some with cubic (presumably) ice crystals (Fig. 9c). Similar premature cleaving and consequent self-contamination is thought to be particularly prone to occur when using certain types of double replica devices that require manipulation and insertion of previ-

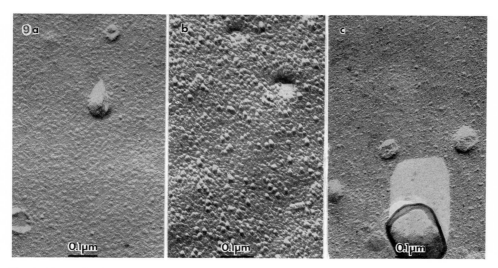

Fig. 9. Contamination patterns observed after premature fracture. **a:** Flat-topped pseudoparticles after accidental partial fracture. ×80,000. **b:** "Cobblestones" after deliberate partial fracture. ×85,000. **c:** Cubic ice crystals deposited between fracture faces formed during freezing or specimen storage. ×75,000. (See R. L. Steere, et al., *this volume.*)

ously frozen specimens into devices with close physical tolerances. (See also R. L. Steere et al., *this volume.*)

SUMMARY

Freeze-fracture specimens cleaved at conventional or ultracold temperatures may become contaminated very rapidly by frozen water vapor. Within seconds, the layer of ice may alter particle size and obliterate relevant structural detail. The primary sources of contamination in the Balzers freeze-etch device are (a) water vapor and glycerol subliming from the chip-coated knife edge, (b) water vapor subliming from adjacent specimens and from the hoar-frosted specimen stage, and (c) residual water vapor subliming from exposed surfaces in the bell jar. Despite the most stringent precautions to avoid contamination from these sources, however, specimens may prefracture during initial freezing or during specimen storage, resulting in severe but unavoidable contamination of apposed surfaces. Thus, each investigator must be able to recognize the various morphologies of water vapor contamination and be prepared to discard a replica or a series of replicas rather than attribute artifactual alterations in membrane structure to the biological phenomenon under investigation.

ACKNOWLEDGMENTS

This work was supported by grant NS 14648 from the National Institutes of Health, and by a grant from the Muscular Dystrophy Association.

REFERENCES

1. Bertaud, W. S., Rayns, D. G., and Simpson, F. O. (1970): Freeze-etch studies on fish skeletal muscle. *J. Cell Sci.,* 6:537–557.
2. Branton, D., Bullivant, S., Gilula, N. B., Karnovsky, M. J., Moor, H., Mühlethaler, K., Northcote, D. H., Packer, L., Satir, B., Satir, P., Speth, V., Staehelin, L. A., Steere, R. L., and Weinstein, R. S. (1975): Freeze-etching nomenclature. *Science,* 190:54–56.
3. Bray, D. F., and Rayns, D. G. (1976): A comparative freeze-etch study of the sarcoplasmic reticulum of avian fast and slow muscle fibers. *J. Ultrastr. Res.,* 57:251–259.
4. Chalcroft, J. P., and Bullivant, S. (1970): An interpretation of liver cell membrane and junction structure based on observation of freeze-fracture replicas of both sides of the fracture. *J. Cell Biol.,* 47:49–60.
5. Crowe, L. M., and Baskin, R. J. (1978): Freeze-fracture of intact sarcotubular membranes. *J. Ultrastr. Res.,* 62:147–154.
6. Deamer, D. W., and Baskin, R. J. (1969): Ultrastructure of sarcoplasmic reticulum preparations. *J. Cell. Biol.,* 42:296–307.
7. Deamer, D. W., Leonard, R., Tardieu, A., and Branton, D. (1970): Lamellar and hexagonal phases visualized by freeze-etching. *Biochim. Biophys. Acta,* 219:47–60.
8. Rash, J. E., and Ellisman, M. H. (1974): Studies of excitable membranes. I. Macromolecular specializations of the neuromuscular junction and the nonjunctional sarcolemma. *J. Cell Biol.,* 63:567–586.
9. Rash, J. E., Ellisman, M. H., Staehelin, L. A., and Porter, K. R. (1974): Molecular specializations of excitable membranes in normal, chronically denervated, and dystrophic muscle fibers. In: *Exploratory Concepts in Muscular Dystrophy. Excerpta Medica Int. Congr. Series No. 333:*271–289.
10. Staehelin, L. A., and Bertaud, W. S. (1971): Temperature and contamination dependent freeze-etch images of frozen water and glycerol solutions. *J. Ultrastr. Res.,* 37:146–168.

Freeze Fracture: Methods, Artifacts, and Interpretations, edited by J. E. Rash and C. S. Hudson. Raven Press, New York © 1979.

Electronically Interlocked Electron Gun Shutter for Preparing Improved Replicas of Freeze-Fracture Specimens

*Mark H. Ellisman and **L. Andrew Staehelin

*Department of Neurosciences, University of California, San Diego, La Jolla, California 92093; and **Department of Molecular, Cellular, and Developmental Biology, University of Colorado, Boulder, Colorado 80309*

After the conventional platinum–carbon (Pt–C) evaporator was replaced by an electron gun evaporator in our Balzers freeze-etch device, we noted that the newer replicas were of a consistent quality but that they were inferior to the best produced with the older resistance-type source. Test specimens consisting of 20% glycerol in water indicated that when the electron gun was used according to the standard operating procedure (gun activation following the final cleavage) some melting and sublimation of the fracture face occurred before replication. Since the heating up of the Pt–C source in the electron gun takes up to 5 sec, compared to less than 1 sec with a resistance heating device, we attributed the specimen changes to heat exposure during the warming up of the electron gun. To test this idea we prepared some replicas of specimens that were not exposed to the heat of the warming up electron gun by delaying the final fracture until 4 to 5 sec after the gun was switched on. Under these conditions the fresh fracture face is replicated immediately after it is formed. The replicas prepared in this manner appeared distinctly "crisper." However, because the procedure was cumbersome, we developed a simple electromagnetic shutter system and electronic control unit to monitor and regulate metal evaporation in the Balzers freeze-etch apparatus. This automatic shutter system allows for: (a) rapid and repeatable coating of specimens within 1 sec after fracturing; (b) better preservation of fractured surfaces by minimizing their heat exposure by the gun; (c) recording of elapsed time of total anode use (and thus how much Pt is left in gun) and the number of seconds each coating takes (in practice a function of the position of the anode tip in the tungsten cathode holder).

DESCRIPTION OF THE SHUTTER AND CONTROL UNIT

The shutter control unit (Fig. 1) is connected to the gun power supply circuit (Balzers control unit EVM 052) at a point at which there is a low voltage

FIGS. 1 and 2. 1: This is a front view of an early model of the shutter control unit, illustrating the counter and the delay time control. **2:** The shutter in position occluding the opening of the electron gun. (Figures courtesy J. E. Rash and W. F. Graham.)

output proportional to the cathode current as indicated on the mA meter. This is a simple connection and requires only the soldering of two wires (a signal and ground) to easily accessed pins within the power supply. The potential across these two points is detected by the input segment of the shutter control unit circuit. The level of signal necessary for detection (triggering) is set by a variable resistance (threshold control knob, not shown). Once threshold is surpassed (operationally about 10 mA), the timing portion of the circuit is activated initiating the onset of the shutter delay interval (i.e., gun warm-up period). This interval, the time between initiation of gun current flow and shutter opening, is another function controlled by a variable resistance. After the specified delay the shutter is opened by activation of the coil of a modified relay to which a nonconductive arm is attached. This movable shutter (Fig. 2, *arrow*) covers the opening of the electron gun heat shield. Upon cessation of current flow from the power supply to the gun (switched off by the interlocked quartz crystal monitor), power to the coil is terminated and the shutter automatically closes. The closing of the shutter in response to the signal from the quartz crystal monitor immediately stops further deposition of Pt–C on the specimen. Without the shutter, crystal monitor values after shadowing are generally 20 to 40 Hz higher than the crystal shut-off value; with the shutter they are never more than 10 Hz higher. This device not only reduces the heat load of the fracture face and ensures replication of the specimen with Pt–C at a maximal rate, but also quickly terminates exposure during the period after the desired thickness has been evaporated.

OPERATION

The sequence of operations for fracturing and replication with the electron gun quartz crystal monitor and shuttering system [with specimens and knife (if used), cooled and ready to fracture] is as follows:

1. With "filament" off, preset voltage of power supply to 1900 V.
2. Turn off function switch of the gun power supply (turning this switch on is the final event activating the gun–crystal–shutter system).
3. Turn filament current knob to a position known (and marked) to correspond to 90 mA on current meter (may be established earlier).
4. Zero (null) quartz crystal monitor by approaching null point turning knob clockwise.
5. Zero (null) elapsed time counter of shutter control unit or record previous elapsed time.
6. Begin cleaving process but *do not make the final cut.* (If using the double-break device, do not break yet.)
7. Switch on power supply by turning the function switch to "HV ON" thereby activating the electron gun.
8. *Quickly* (within about 3 sec), as gun heats and before shutter opens, either make the final cut or break the specimen with the double-replica device. (Time this manuever so that the knife clears the specimen and crystal monitor just as the shutter opens.)
9. The elapsed time for metal evaporation is regulated by the quartz crystal monitor interlocked to the power supply. The cessation of power to the gun simultaneously closes the shutter.
10. Reinforce replica with carbon and process as usual.

The device described above has proved useful and reliable for over 4 years. Other shutter systems have been constructed more recently with more modern electronic circuitry but all perform the same essential functions. Some alterations in the actual shutter solenoid design illustrated here are anticipated to accommodate the newer variable-angle shadowing system. Details of construction of the shutter system may be obtained from Dr. Ellisman.

ACKNOWLEDGMENTS

This work was supported by research grants to L. Andrew Staehelin from N.I.G.M.S. No. 18639 and to Mark H. Ellisman from N.I.N.C.D.S. No. NS14718 and the Muscular Dystrophy Association of America. Important contributions in refinement of the shutter circuit were also made by Paul Lutis, Ed Cole, and Steve Young.

Freeze Fracture: Methods, Artifacts, and Interpretations, edited by J. E. Rash and C. S. Hudson. Raven Press, New York © 1979.

Advances in Ultrahigh Vacuum Freeze Fracturing at Very Low Specimen Temperature

Heinz Gross

Institut für Zellbiologie, Eidgenössische Technische Hochschule, Hönggerberg, 8093 Zürich, Switzerland

Freeze etching or freeze fracturing has become established as an important technique for ultrastructural investigations. A characteristic feature of freeze fracturing of biological specimens is membrane cleavage, which leads to the visualization of the so-called intramembrane particles on a smooth background (2,5,17). Conventional freeze fracturing is carried out in a vacuum of $\sim 10^{-6}$ torr and at a specimen temperature of $\sim -100°C$. If the structural details of both fracture faces produced by the cleavage are precisely compared, astonishing morphological differences of complementary sites are usually observed (5,8,17). The lack of complementarity in membrane fracture faces, and particularly the absence of E-face depressions corresponding to the P-face particles, calls into question the reliability of the structural record. Each of the preparational steps involved in the technique possesses its particular problems that may influence the apparent structure of the object being examined.

The task of freezing is to solidify a specimen in the natural state without any changes in structure and chemistry. The main problems derive from segregation phenomena in the liquid phases of cells, that is, in the cytosol and in membranes (16,17,25).

In fracturing a mechanical separation of membrane components takes place, producing heat that in turn facilitates plastic deformation. At the usual temperature of $-100°C$ drastic plastic deformation has been shown in some specimens (5,17,20). Although our understanding of the fracturing process is limited, it stands to reason that deformation can be lessened by fracturing at very low temperatures.

During the time between cleaving and coating there exists the danger of contaminating the fracture faces with residual gases of the vacuum. Gas molecules impinging on a solid-state surface can be retained by adsorption or by condensation. A gas *condenses* only if its partial pressure is greater than its saturation vapor pressure at the temperature of the surface. Molecules can, however, *absorb* to the surface even in the unsaturated condition up to a certain limiting surface coverage. Contamination of the fracture faces by condensation can be prevented by the choice of specimen temperature for a given vacuum.

Adsorption cannot be prevented, and all impinging gas molecules are retained in at least one monomolecular layer. Figure 1 shows the vapor pressure curves (13) for most of the gases found in vacuum systems in the lower pressure region. At − 100°C all saturation vapor pressures are greater than 10^{-6} torr, and the pressure of water vapor, the major residual gas in conventional high vacuum systems (8,17), is 10^{-5} torr. If the temperature is lowered to − 150°C the saturation vapor pressure of water is reduced to about 10^{-11} torr, far below the partial pressure of water vapor in conventional vacuum systems. This means, inevitably, contamination of the fracture faces by condensing water molecules. If experiments are to be performed at very low specimen temperature, this unwanted coating of water can be limited only by reducing the deposition rate of the impinging water molecules.

Finally, the fidelity of the structural record depends on the amount of damage introduced by surface heating during replication and the granularity (grain size, grain density) of the heavy metal coat (7,17,25). The amount of heat damage caused by the heat of condensation and thermal radiation of the hot sources (at least 2,500°C) cannot be ascertained directly. No means are available to measure or even to estimate the temperature at the surface of a cooled object exposed to an energy flow of this kind. We can only make the not unreasonable assumption that lowering of the specimen temperature should also lower the actual surface temperature and thereby reduce any damage of the heat-sensitive specimens.

All these factors, except for segregation phenomena during freezing, might be responsible for the lack of complementarity in membrane fracture faces. All of them are temperature dependent, and at lower specimen temperature, less distortion of fractured structures, less damage of radiation-heated surfaces, and less surface diffusion of condensing heavy metal atoms can be expected (7,8). The only drawback is the increased contamination at low temperatures. If all impinging residual gas molecules (mainly water vapor) are retained, the

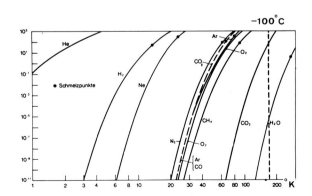

FIG. 1. Vapor pressure curves of the residual gas constituents. (Schmelzpunkte = melting points.)

freshly cleaved specimen surfaces would be covered with about one monomolecular layer per second (assuming homogeneous distribution) under conventional high-vacuum conditions (10^{-6} torr). The deposition rate is reduced 1,000 times if a vacuum of 10^{-9} torr, that is, ultrahigh vacuum (UHV), is applied. In other words, if it is possible to carry out freeze fracturing under UHV, practically contamination-free replication are possible independent of the specimen temperature.

The problems of reconciling UHV and freeze fracturing are as follows: Conventionally the cold specimens are affixed to a precooled stage under atmospheric conditions; this means hoar-frosting of the specimens and their surroundings. Ice crystals with poor thermal contact to the cooled surfaces, deposited very near the resulting fracture faces, represent a nonmeasurable source of water vapor contamination similar to the specimen chips left on the knife edge after fracturing with the microtome assembly (21). A second problem derives from gas production by the coating procedure. In standard freeze fracturing no special precautions are taken to prevent a pressure increase during evaporation. If the same evaporators are operated under UHV, a drastic vacuum breakdown takes place (to at least 10^{-6} torr) before any platinum atom has condensed on the fracture faces. Considering these requirements, we have developed a prototype UHV-freeze-fracture apparatus that allows fracturing and shadowing of specimens down to $-196°C$, while maintaining a vacuum of 10^{-9} torr (7,8). It consists of a modified Balzers-UHV unit, equipped with an airlock system that enables the introduction of nonhoar-frosted specimens directly into the evacuated vacuum chamber. At present this apparatus enables us to perform three UHV experiments per day. After three experiments the vacuum chamber must be vented in order to exchange the carbon evaporator. A further feature is the automatic control of Pt/C shadowing and C-backing. The deposition rate is monitored before opening a pneumatically driven shutter. The shutter is coupled to the quartz thin film monitor. When the desired film thickness is reached, the shutter closes automatically. Exact measurement and control of the film thickness, the deposition rate, and the specimen temperature are important because these parameters influence the granularity of the shadowing films, and hence the resolution of the replica.

RESULTS AND DISCUSSION

Assessment of methodological progress is greatly facilitated if methods of image averaging (i.e., optical diffraction, digital image reconstruction) can be applied. These techniques provide information about the average repeating features present in periodic structures. The plasmalemma of starved baker's yeast cells, which contains a suitable periodic structure, was chosen as a first test specimen.

After conventional freeze fracturing (10^{-6} torr, $-100°C$), the fracture faces of the yeast plasmalemma show the known astonishing morphological differences between complementary sites described earlier (4,8,17). The P-face exhibits patches of hexagonally arranged particles whereas the E-face contains at best a few depressions but never enough to suggest complementarity of the two faces. The structural record after UHV freeze fracturing at $-196°C$ is shown in Figs. 2a (P-face) and 2b (E-face). Between the most prominent structures of the yeast plasmalemma, that is, the trough-like invaginations, patches with paracrystalline structure are visible. On the P-face hexagonally arranged particles with a volcano-like shape and on the E-face corresponding ring-like depressions can be seen, but their regularity seems less well defined than on the P-face. Complementarity of these periodic features can be demonstrated by digital image filtration with the results shown in Figs. 2c and d (14). The volcano-like particles on the P-face have their obvious counterparts in the ring-like depressions on the E-face. We call this paracrystalline structure with 165 Å lattice repeats the "main structure." This main structure is the only feature that might have been expected from conventional freeze-fracturing experiments, although correspondence of P- and E-faces had not been established conclusively. On the E-face an additional feature that we call the "substructure," is clearly revealed

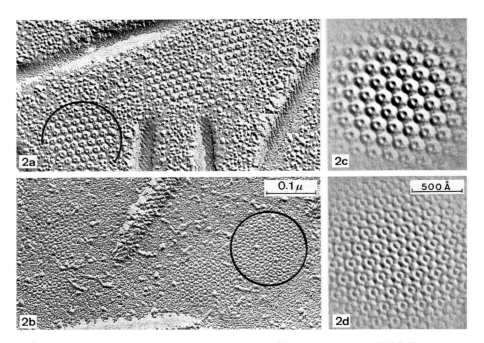

FIG. 2. Yeast plasmalemma, freeze fractured at $-196°C$ under $1 \cdot 10^{-9}$ torr (UHV). The hexagonally ordered structures (encircled) are seen to consist of volcano-like particles on the P-face **(a)** and corresponding ring-like depressions on the E-face **(b)**. a and b: ×44,000. Complementarity of periodic features on the P- and E-face is revealed by digital image filtration **(c and d)**. c and d: ×264,000.

(Fig. 2d). Each ring of the main structure is surrounded by six pits. On the P-face regular features appear (Fig. 2c), in addition to the main structure. However, these particle-like structures do not seem to match exactly with the E-face pits (14).

The improvement attained with the new technique can be attributed primarily to reduced plastic deformation during fracturing, and secondarily to reduced heat damage during replication at lower specimen temperature (7,8). The structural record is considered reliable. The nearly perfect complementarity leaves little doubt as to the reality of the structures derived by image averaging. It became apparent that other test specimens, biochemically and structurally better defined than the rather complex plasmalemma of yeast, would be needed to learn how to interpret the structural record and assess improvements. Such a test specimen should meet the following criteria: (a) its structure should be known from independent measurements; (b) it should contain features sufficiently fine that improved resolution of structural details on the replica is clearly apparent; (c) there should be standard procedures for reproducibly preparing defined specimens; and (d) the specimen should preferably have a simple biochemical structure. The purple membrane of *Halobacterium halobium* (23) appeared a suitable candidate. It is biochemically rather simple, easy to obtain, and standard conditions for culture have been defined (1,18). The purple membrane is a specialized part of the plasmalemma of *Halobacterium halobium* and contains only one hexagonally arranged protein (bacteriorhodopsin) (18). The structure of the protein has been described in three dimensions at 7 Å resolution by computer analysis of electron diffraction patterns and corresponding micrographs (12,24).

Conventional freeze-fractured preparations of whole cells show the appearance typical of most freeze-fractured cell membranes. The P-face exhibits quasi-crystalline aggregates of particles whereas the smooth E-face is characterized by a lack of corresponding depressions (1,11,15), but in exceptional cases indications of a regular structure have been observed (1,11). The results of UHV freeze fracturing at $-196°C$ are shown in Figs. 3a–c. Periodic structures of equal lattice constant were found on both fracture faces (11,15). The P-faces (Figs. 3a and b) reveal preferential localization of the condensed shadowing material. Larger amounts of Pt/C (~ 15 Å) lead to a clustered appearance (Fig. 3a). The clusters are apparently formed by accumulation in special nucleation sites (decoration) and not by coalescence of randomly distributed high-density nuclei. This becomes apparent with smaller amounts of Pt/C (~ 6 Å), (Fig. 3b) which show crystallites of much smaller size but similar center-to-center spacing. On the E-face the same amount of deposited Pt/C (6 Å) produces finer and smaller spaced grains and ring-like structures become visible. The periodic signal component in both the P- and the E-face was extracted by digital image processing. The periodic component of the P-face looks similar to filtered images of conventional replicas (11,15), showing only the basic lattice period and particulate structures (Fig. 3d). The processed versions obtained with different amounts of Pt/C have very similar appearances, in contrast to the impression given by

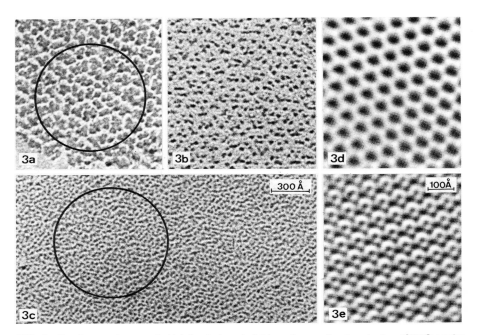

FIG. 3. Purple membrane of *Halobacterium halobium*, freeze fractured at −196°C under 1 · 10⁻⁹ torr (UHV). Large amounts (~ 15 Å) of Pt/C lead to a clustered appearance on the P-face **(a)**. That migration and accumulation occur preferentially at certain sites (decoration) is revealed by smaller amounts (~ 6 Å) of Pt/C **(b)**; finer crystals but a similar center-to-center spacing are seen. On the E-face, 6 Å Pt/C produces finer and more closely spaced crystals **(c)**. a–c: ×348,000. The periodic component of the P-face **(d)** shows only the basic lattice period and particle-like features, whereas the E-face, in addition, shows internal structure, consisting of ring-like depressions **(e)**. **d** and **e:** ×696,000.

the original micrographs. The size of aggregates of Pt/C is thus of secondary importance. The periodic component on the E-face shows an internal structure consisting of ring-like depressions (Fig. 3e). The outer and inner diameters of these depressions are consistent with the model of Unwin and Henderson (12,24) which postulates ring-like arrangement of intrinsic proteins surrounded by lipids in bilayer configuration. The space in the middle of the ring (diameter ~ 20 Å) is also thought to be filled with lipid (12,24).

The remaining inconsistency could be caused by residual plastic deformation during fracturing even at very low temperature and/or by unsufficient resolution of Pt/C shadowing on the P-face (11). If the protein stays entirely on the P-face, the protein molecules arranged in a ring may collapse toward each other like fingers in the fracturing process. The results of Pt/C shadowing on the P-face demonstrate that at higher resolution decoration phenomena can no longer be neglected. In vacuum evaporation, the formation of a thin film proceeds via thermal accommodation, nucleation, and crystal growth (7,10).

The small amount of deposited metal which is used for shadowing leads to a discontinuous film consisting of small (10 to 50 Å) individual crystals. The formation of individual, more or less evenly spaced, crystals originates from the lateral motion of the condensing atoms, which occurs even at very low specimen temperature. Both the similar appearance of filtered P-face images obtained with different amounts of Pt/C (i.e., smaller and larger crystallites) and the lack of a shadowing direction (Fig. 3d) support the assumption that migration of the shadowing material and nucleation and crystal growth occur at preferential sites.

It is known that grain size and grain location depend not only on the properties of the condensing material, but also on the physicochemical properties of the exposed surface. In other words, the preferential location of crystallites of suitable condensing materials open up the possibility of using condensing materials as labels for regions on the fracture faces with certain physicochemical properties. If the exposed structures cause an unequal distribution of surface forces, the migrating atoms or molecules are able to settle at places with high binding energy where nucleation and hence crystal growth is more probable. Of course, such investigations require surfaces devoid of contamination. Every adsorbed gas molecule alters the distribution of the surface forces and, therefore, also the nucleation process. At a residual gas pressure of $1 \cdot 10^{-9}$ torr (UHV) the formation of a monomolecular layer of contamination requires at least 1,000 sec. Therefore, an undisturbed interaction between the decorating material and the exposed structures is possible. Because the condensation pattern is also strongly influenced by film thickness, deposition rate, surface temperature, and the vacuum conditions, our UHV device was so constructed that all of these parameters can be measured and monitored (7,8).

As mentioned above, low temperature ($\leq -110°C$) surfaces become contaminated under conventional vacuum conditions ($\sim 10^{-6}$ torr) mainly by condensing water vapor. With freeze-fractured lipid–water and glycerol–water mixtures, Deamer et al. (6) and Staehelin and Bertaud (21) have shown that the deposition of residual gases is highly substrate specific and that artificial structures are formed that resemble those found on natural membranes.

To examine these decoration and contamination phenomena, clean membrane fracture faces were exposed to pure water vapor under controlled conditions. For this a device was developed that allows the production of pure water vapor (9,10). A small container filled with copper sulfate pentahydrate is heated under vacuum to 100°C; the water of hydration thereby released enters a storage vessel. From here water vapor is released through a special valve into the UHV chamber. The purity of the released water vapor is at least 99%, as shown by mass spectroscopy (9,10).

To achieve precise and reproducible control of the exposure, the required partial pressure of water was established prior to producing the fracture faces. At $1 \cdot 10^{-9}$ torr the inlet valve for water vapor was opened and regulated until the required partial pressure was attained. Fracturing was then performed at

— 196°C and the resulting fracture faces exposed for a specified time. Then the inlet valve was closed, causing immediate return to UHV of at least 3 × 10^{-9} torr. Subsequent coating was performed by Pt/C-shadowing and C-backing.

Figs. 4a to c show the effect of controlled exposure of yeast plasmalemma E-faces to different amounts of pure water vapor. The micrographs demonstrate that even at the very low specimen temperature of — 196°C water does not condense uniformly as a homogeneous layer. On the E-face exposed for 60 sec to 5 × 10^{-9} torr (Fig. 4a) the paracrystalline regions (encircled) appear with about the same distinctness as under best UHV conditions (Fig. 2b). The only indication of water condensation are some more or less statistically distributed particulate structures *(arrows)*. On the E-face exposed for 60 sec to 1 · 10^{-6} torr all topographic details of the depressions are hidden (Fig. 4c). However, the location of the paracrystalline regions (encircled) can be recognized because particulate structures have been formed in patches of the same size and location. Discrete ice crystals with a cubic shape *(arrows)* can be detected in these patches. Such individual ice crystals can also be found on the E-face exposed for 60 sec to 5 × 10^{-8} torr (Fib. 4b). In the paracrystalline regions (encircled), small discrete ice crystals and nonhidden ring-like depressions can be recognized. These observations imply that the lattice of the underlying structures is reflected in the arrangement of the ice crystals ("specific decoration"). This assumption can be verified by the use of optical diffraction and digital image processing (see also refs. 9 and 10). In Fig. 4d the filtered image of a paracrystalline region after UHV freeze fracturing at — 196°C without additional water vapor condensation is repeated. Figure 4e shows the filtered image of a paracrystalline region exposed 5 sec to a water pressure of 10^{-5} torr. The comparison of the two pictures demonstrates that the ring-like depressions that correspond to the volcano-like P-face particles (main structure) are predominantly decorated with ice crystals. Precise inspection reveals small ice crystals over the pits of the substructure also. The formation of discrete ice crystals requires lateral mobility of the condensing water molecules. The preferred formation of ice crystals on the E-face depressions must result from a higher binding energy on the surface of these structures than on the surface of the surrounding areas.

Freeze fracturing under UHV-conditions with and without controlled exposure to pure water vapor, the major residual gas in conventional high vacuum systems, also allows us to describe the structural alterations introduced by condensing

FIG. 4. E-faces of yeast plasmalemma, freeze fractured at — 196°C and prior to replication exposed for 60 sec to increasing partial pressures of pure water vapor: **(a)** 5 × 10^{-9} torr; **(b)** 5 × 10^{-8} torr; **(c)** 1 · 10^{-6} torr (background vacuum 1 · 10^{-9} torr). The micrographs demonstrate that even at — 196°C water does not condense uniformly. Particle-like ice crystals are formed mainly in the paracrystalline regions (encircled). The filtered images of two ordered regions with **(e)** and without **(d)** condensed water vapor reveal that the ring-like depressions, which correspond to the volcano-like P-face particles, are "specifically decorated" with ice crystals. **a–c:** ×150,000; **d** and **e:** ×250,000.

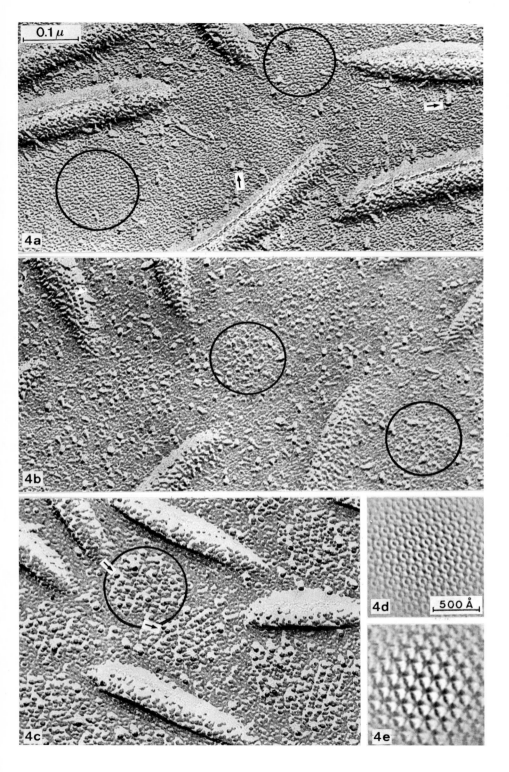

water molecules (main contamination). In UHV experiments without additional water condensation, patches of depressions practically devoid of particulate structural features were found on the E-face [Fig. 2b (10)]. By increasing the amount of condensed water vapor the number of ice crystals grew and reached a density at which all topographic details were hidden. The actual uncontaminated structure on the E-face seems therefore to consist of depressions without any particulate structures. Both the appearance of water condensation as discrete, particle-like ice crystals and their preferred growth on the depressions opposite to the corresponding actual P-face particles may easily lead to misinterpretation of ice crystals as membrane structures. Therefore the structural features of membrane fracture faces produced under high-vacuum conditions ($\sim 10^{-6}$ torr) at low specimen temperature ($\leq -110°C$) have to be interpreted with caution. Furthermore, the particle-like condensation of water vapor, even at the very low specimen temperature of $-196°C$, and the simultaneous occurrence of holes and ice crystals on the same fracture face, disprove the widely held opinion that portrayal of holes in membrane fracture faces is indicative of a contamination-free replication (20).

It should be pointed out that, even in pictures obtained with UHV freeze fracturing, a few ice crystals can be found. A possible source of the water molecules that form ice crystals even under UHV conditions may be the small outbursts of water vapor observed during fracturing as shown in Fig. 5. The course of the total gas pressure recorded during successive fracturing of three specimens shows that the pressure rises persist less than 1 sec and do not surpass $1 \cdot 10^{-7}$ torr. The simultaneously recorded mass spectrogram, representing exclusively the main peak of water, proves that the pressure rises are caused by water vapor.

A better understanding of sources of contamination is essential for improving freeze fracturing under conventional high-vacuum conditions also. We must recall that in such systems there exist, in addition to the residual gases inside the vacuum chamber, further sources of contamination, such as hoar-frost formation on the specimen and its surroundings and gas production by the coating procedure, that have been eliminated in our UHV device (7,8). For this reason the precautions most widely taken to prevent deposition of condensable gases at low specimen temperatures, such as shadowing immediately after fracture, protection of the fracture faces by the cooled knife at $-196°C$ (Balzers device), or complete enclosure of the specimen in a liquid nitrogen-cooled shroud (19,22) may not always be sufficient to provide clean images. This pessimistic assessment is supported by published E-face pictures on which ice crystals are clearly revealed (e.g., Figs. 9 and 10 in ref. 20, Fig. 13 in ref. 4, and Fig. 2a in ref. 3).

Maintaining UHV conditions during the whole freeze-fracturing process is the surest way to get clean pictures routinely. Such reliability is especially important when no structure determination by independent methods exists and when the specimens to be examined are hard to obtain or involve lengthy preparations. With present-day vacuum systems, equipped with suitable airlocks for both

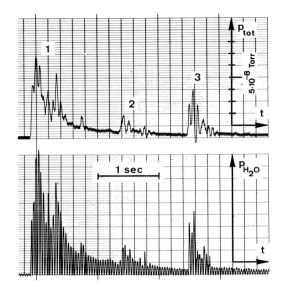

FIG. 5. Outbursts of water vapor recorded during successive fracturing of three specimens.

the specimens and the evaporators, routine freeze fracturing under UHV conditions is possible with running times on the order of 1 hr.

CONCLUSIONS

By fracturing at the very low specimen temperature of $-196°C$ under UHV conditions we have achieved improved topographic resolution and improved complementarity of the fracture faces with both specimens, the plasmalemma of starved yeast cells and the purple membrane of *Halobacterium halobium.* Depressions corresponding to the P-face particles are clearly revealed on the E-face of both membranes. A comparison of the results obtained for the purple membrane compared with the results obtained with the three-dimensional model of Henderson and Unwin (12) in which ring-like arrangement of the intrinsic protein is postulated, indicates that the course of the fracture goes around the protein complexes leaving all of the protein on the P-face. For the main structure of the yeast plasmalemma the same interpretation seems to be valid (10). The observation that the volcano-like P-face particles are elevated by at least 40 Å (14) suggests that the particles penetrate the outer half of the plasmalemma and reach the outer membrane surface. This assumption is supported by the fact that these depressions are obviously decorated with ice crystals (Fig. 4e) (10). This would be expected if at these places the outer leaflet of the lipid bilayer is interrupted, exposing a surface area with polar properties. Consistent with this interpretation is the continued visibility of the corresponding P-face particles even after heavy water deposition (10). Intrinsic membrane structures are expected to be hydrophobic and hence to interact only weakly with polar water molecules. We are aware that this speculative interpretation must be confirmed with the help of simple model systems. However, these initial experiments do show that with controlled deposition of suitable condensing gases on clean surfaces, "specific decoration" is possible even on complex membrane fracture faces and therefore can be used in combination with UHV freeze fracturing to label regions with distinct physicochemical properties. Such combined investigations may be useful in relating structure and function of the recorded intrinsic membrane components.

By further optimizing the different experimental parameters and by applying more suitable shadowing materials we feel certain to obtain reliable structural details at a resolution of 15 to 10 Å for image averaged periodic specimens.

ACKNOWLEDGMENTS

I am grateful to Prfs. H. Moor and E. Bas for continued support and stimulating discussions. I also wish to thank Dr. O. Kübler for making the digital filtered images, Dr. U. Müller for supplying the *Halobacterium halobium* cells, Miss D. Walzthöny and Mr. A. Frey for their excellent technical assistance, and Dr. D. Turner for his critical reading of the manuscript and assistance with English.

These studies were supported by Swiss National Science Foundation, grant No. 3.690–0.76.

REFERENCES

1. Blaurock, A., and Stoeckenius, W. (1971): Structure of the purple membrane. *Nature, (New Biol.),* 233:152–155.
2. Branton, D. (1971): Freeze-etching studies of membrane structure. *Philos. Trans. R. Soc. Lon. (Ser. B),* 261:133–138.
3. Branton, D., and Kirchanski, S. (1977): Interpreting the results of freeze-etching. *J. Microsc.,* 111:117–124.
4. Bullivant, S. (1973): Freeze-etching and freeze-fracturing. In: *Advanced Techniques in Biological Electron Microscopy,* edited by J. K. Koehler, pp. 67–112. Springer-Verlag, Berlin, Germany.
5. Bullivant, S. (1974): Membranes. Freeze-etching techniques applied to biological membranes. *Philos. Trans. R. Soc. Lon. (Ser. B),* 268:5–14.
6. Deamer, D. W., Leonard, R., Tardieu, A., and Branton, D. (1970): Lamellar and hexagonal lipid faces visualized by freeze-etching. *Biochim. Biophys. Acta,* 219:47–60.
7. Gross, H. (1977): Gefrierätzung im Ultrahochvakuum (UHV) bei − 196°C. Ph.D. Dissertation 5881, ETH, Zürich, Switzerland.
8. Gross, H., Bas, E., and Moor, H. (1978): Freeze-fracturing in ultrahigh vacuum (UHV) at − 196°C. *J. Cell biol.,* 76:712–728.
9. Gross, H., and Moor, H. (1978): Decoration of specific sites on freeze-fractured membranes. *Proc. Int. Congr. Electron Microsc., 9th.,* 2:140–141.
10. Gross, H., Kübler, O., Bas, E., and Moor, H. (1978): Decoration of specific sites on freeze-fractured membranes. *J. Cell Biol.,* 79:646–656.
11. Gross, H., Kübler, O., and Moor, H. (1979): *J. Cell biol., (manuscript submitted for publication.)*
12. Henderson, R., and Unwin, P. N. T. (1975): Three-dimensional model of purple membrane obtained by electron microscopy. *Nature (Lond.),* 257:28–32.
13. Honig, R. E., and Hook, H. O. (1960): *RCA Review.* 21:3.

14. Kübler, O., Gross, H., and Moor, H. (1978): Complementary structures of membrane fracture faces obtained by ultrahigh vacuum freeze-fracturing at − 196°C and digital image processing. *Ultramicroscopy,* 3:161–168.
15. Kübler, O., and Gross, H. (1978): UHV freeze-fracturing and image processing applied to the purple membrane. *Proc. Int. Congr. Electron Microsc., 9th.,* 2:142–143.
16. Moor, H. (1964): Die Gefrier-Fixation lebender Zellen und ihre Anwendung in der Elektronenmikroskopie. *Z. Zellforsch. Mikrosk. Anat.,* 62:546–580.
17. Moor, H. (1971): Recent progress in the freeze-etching technique. *Philos. Trans. R. Soc. Lon. (Ser. B),* 261:121–131.
18. Oesterhelt, D., and Stoeckenius, W. (1971): Rhodopsin-like protein from the purple membrane of *Halobacterium halobium. Nat. New Biol.,* 233:149–152.
19. Sleytr, U. B., and Umrath, W. (1976): Freeze-etching: Technical developments and general interpretation problems. *Proc. Eur. Conf. Electron Microsc., 6th.* 2:50–55.
20. Sleytr, U. B., and Robards, A. W. (1977): Freeze-fracturing: A review of methods and results. *J. Microsc.,* 111:77–100.
21. Staehelin, L. A., and Bertaud, W. S. (1971): Temperature and contamination dependent freeze-etch images of frozen water and glycerol solutions. *J. Ultrastruct. Res.,* 37:146–168.
22. Steere, R. L. (1973): Preparation of high resolution freeze-etch, freeze-fracture, frozen-surface, and freeze-dried replicas in a single freeze-etch module, and the use of stereo electron microscopy to obtain maximum information from them. In: *Freeze-Etching, Techniques and Applications,* edited by E. L. Benedetti and P. Favard, pp. 223–255. Société Française de Microscopie Electronique, Paris.
23. Stoeckenius, W., and Kunau, W. H. (1968): Further characterization of particulate fractions from lysed cell envelopes of *Halobacterium halobium* and isolation of gas vacuole membranes. *J. Cell Biol.,* 38:337–357.
24. Unwin, P. N. T., and Henderson, R. (1975): Molecular structure determination by electron microscopy of unstained crystalline specimens. *J. Mol. Biol.,* 94:425–440.
25. Zingsheim, H. P. (1972): Membrane structure and electron microscopy. The significance of physical problems and techniques. (Freeze-etching). *Biochim. Biophys. Acta,* 265:339–366.

Freeze Fracture: Methods, Artifacts, and Interpretations, edited by J. E. Rash and C. S. Hudson. Raven Press, New York © 1979.

Fine Structure of Yeast Plasma Membrane After Freeze Fracturing in a Simple Shielded Device

Stanley Bullivant, Peter Metcalf, and Kennedy P. Warne

Department of Cell Biology, University of Auckland, Auckland, New Zealand

Recently, Gross et al (6) claimed that it was necessary to use an expensive and complex ultrahigh-vacuum freeze-fracture apparatus to demonstrate the fine structure of the paracrystalline arrays of the yeast plasma membrane, namely the volcano-like particles of the P-face and the ring-like depressions of the E-face. Subsequently, it was shown that these structures could be seen after preparation in a standard, but relatively complex, commercially available apparatus equipped with good anticontamination shielding (10). Some years ago McNutt and Weinstein (8) demonstrated the E-face rings after preparation in a simple freeze-fracture apparatus with integral cold shielding, and indeed they proposed that the visibility of the rings be used as a criterion for replica quality. We have used the same apparatus (that of Bullivant and Ames, ref. 3) and have now shown that the P-face volcanoes and E-face rings are easily seen, and, in addition, that details of the E-face previously only elucidated clearly by averaging (7) can be seen easily on the micrographs. This demonstrates, in agreement with Sleytr and Messner (10), that adequate anticontamination shielding of the specimen, rather than ultrahigh vacuum in the whole apparatus, allows the preservation of fine detail. We also show that the P-face structure of the paracrystalline arrays is dependent on the way the yeast fractures.

MATERIALS AND METHODS

All replicas were made using the simple metal block apparatus with shadowing tunnels introduced by Bullivant and Ames (3) and described in detail by Bullivant (1).

Baker's yeast *(Saccharomyces cerevisiae)* was obtained locally as a compressed cake (Dominion Yeast Co., Auckland). The yeast was either fractured directly from the cake, or alternatively after suspending in water at room temperature for 1 hr and gently centrifuging into a pellet. Fracturing was done by two different methods. In the first, the yeast was placed in a 1.7 mm internal diameter polyethylene tube that fitted snugly in the specimen holder. The tube was pre-nicked around its circumference so that when frozen it could be broken rather easily at the line of the nick by a gentle sideways tap. In the second method, the trough of the complementary replica holder (4) was filled with the yeast and the frozen preparation broken by applying a bending moment (torque) to the two parts. It was harder to make the break than with the prenicked plastic tube. To obtain complementary replicas the method was modified by the addition of a pair of in-register gold London Finder grids clamped between the two parts of the holder (9).

After freezing in liquid nitrogen, the specimen is broken under the same liquid and placed into the vacuum evaporator while still covered with liquid nitrogen and enclosed in a three-tier brass block at liquid nitrogen temperature. The construction of the block is such that contaminants from the main volume of the bell jar are prevented from reaching the cold fracture surface because the enclosing blocks are close-fitting and cold. The specimen, in its holder, is within a small interior chamber (Fig. 1). At the start of the experiment, both specimen and blocks are at liquid nitrogen temperature ($- 196°C$). Usually the lid is lifted and shadowing done through the tunnels after 20 min. At that time, the vacuum in the main part of the bell jar is 10^{-5} torr, and above the fracture face it can be assumed that the partial vapor pressure of contaminants is very low. At the time of shadowing, in a typical experiment, the temperature of the lower block is $- 140°C$ and that of the middle block is $- 190°C$. This temperature differential results from the main heat flow being conducted via the base into the lower block. It is not possible to measure the actual surface temperature of the fractured specimen, but it must be between $- 140$ and $- 190°C$. Hence the metal surface above and around the specimen is at a lower

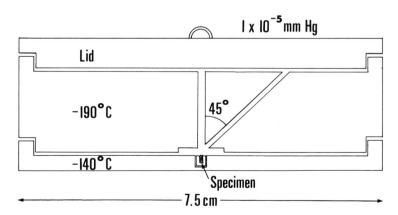

FIG. 1. Cross-sectional diagram of freeze-fracture apparatus. The temperatures of the tunnel and specimen blocks and the evaporator pressure just prior to replication are shown. The purpose of the diagram is to emphasize that the specimen is protected from contamination by being enclosed in the cold interior of the system of blocks. The lid is lifted to allow shadowing through the tunnels. Even at this point the specimen is protected from contamination coming from the main part of the bell jar by the cryo-trapping effect of the cold and narrow tunnels.

temperature than the specimen itself, a necessary condition for trapping any residual contaminants.

A standard Kinney model KSE-2 vacuum evaporator was used. It was equipped with a Ladd dual-electrode set for evaporation of platinum–carbon and carbon by resistive heating.

The replicas were cleaned by floating on sulfuric acid–dichromate solution, washed in water, and picked up on carbon–collodion coated grids. Micrographs of the replicas were made using a Philips EM301 electron microscope.

Diffraction patterns were made from previously selected micrographs using a simple diffractometer constructed with parts from a Spindler and Hoyer 062 micro optical bench kit and a Spectra Physics 165 Argon ion laser.

RESULTS AND DISCUSSION

Structure of P-Faces

It was noticed that after fracturing of cake yeast using the prenicked-tube method, often more than 90% of the cells showed fractures following the plasma membrane. Using the complementary holder method of fracturing on yeast that had been suspended in water and centrifuged, the proportion of membrane fractures decreased and quite often cytoplasmic cross-fractures predominated. However, it was still usually possible to find areas in the replicas where the majority of the cells showed membrane fractures.

We found a reasonably clear-cut difference related to the mode of fracture. Volcano-like particles with distinct craters (Figs. 2a and 3) were found on the P-face in the paracrystalline arrays in areas of the replica where the majority of the yeast exhibited membrane fractures. Diffractograms of such P-face arrays yielded spots out to the fifth order (Fig. 2b). In other areas with more cross-fractures, the particles lacked craters, had a more irregular shape, and often showed signs of plastic deformation (Figs. 4 and 8b).

This association between mode of fracture and appearance of the P-face particles is understandable. The fracture proceeds along a path of average least work (2), and a high proportion of plasma membrane fractures indicates that the membrane path is the easiest one, with less energy being dissipated in deformation. Conversely, a high proportion of cross-fractures indicates that the membrane path involves greater work with more energy being dissipated in deformation. The two types of fracture area (predominantly membrane or predominantly cross) were often found in the same replica. This would indicate that local differences in the fracture might play a part. Fracture is initiated by bending, and both compressive and extensive forces act. It would be interesting to see if, for example, an initial compressive force were associated with a particular type of fracture.

Kübler et al (7) commented that the fracture almost always followed the plasmalemma if highly concentrated suspensions of yeast cells were used. They

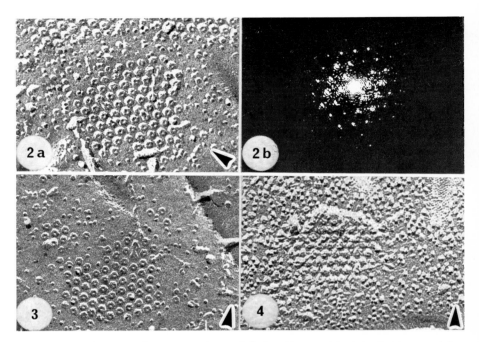

FIGS. 2–4. 2a: P-face array from preparation made by breaking prenicked plastic tube containing yeast frozen directly from the cake. Regular volcano-like particles with craters can be seen. **2b:** Diffraction pattern obtained from the array in **a.** Spots out to the fifth order are visible. **3:** P-face array from same preparation. The local shadow angle is higher and the individual particles have a hexagonal appearance. **4:** P-face array from same preparation made by suspending in distilled water and centrifuging before freezing and fracturing in a complementary holder. The particles have an irregular appearance. Figs. 2–4: *broad arrow, shadowing direction; magnification,* ×174,000.

did not mention any correlation between this type of fracture and the appearance of volcano-like particles on the P-face.

Sleytr and Messner (10) observed that the regular volcano-like P-face particles were generally found in "starved" yeast, whereas the more irregular ones lacking craters were found in "resting" yeast, in contrast to the claim of Gross, et al (6) that these differences resulted from contamination or deformation. We find that the differences can be correlated primarily with the mode of fracture, although this in itself may be related to the physiological state of the yeast.

Structure of E-Faces

The appearance of the paracrystalline arrays on the E-face was not so dependent on the mode of fracture. In almost all replicas examined, two types of E-face appearance and gradations between them were observed.

The first type consisted of a two-part array: (a) a hexagonal pattern of ring-like depressions with the same 16.5 nm center-to-center spacing found for the P-face volcanoes, and (b) small depressions at all the trigonal points of the above hexagonal lattice (Figs. 5, 6, 7a and 8a). Diffractograms of such E-face arrays yielded spots out to the fifth order (Fig. 7b). The trigonal depressions can be seen clearly by direct inspection of the micrographs. (Shadowing direction is indicated by a broad arrow.) They are somewhat more difficult to see in the replicas produced by Gross, et al (6) in ultrahigh vacuum and are only brought out clearly by digital image processing of the micrographs of Gross, et al by Kübler et al (7).

In the second type (not illustrated) the array of E-face rings and depressions was covered with an array of particles similar to those which Gross and Moor (5) believe to be ice crystals condensed in a specific manner on the substrate array. Although this may be certainly so in some instances, there are indications that E-face particles within the paracrystalline arrays are not always the result of contamination, but may be real structures that have stayed with the E-face on fracturing. For example, the stretched particles on the E-face near the edge of the membrane fracture face in Fig. 6 give the appearance of plastic deformation, being very similar in alignment to the deformed P-face particles, which are more commonly seen (11). In the same micrograph there is a neighboring array with virtually no particles on it. It seems unlikely that variations in condensation could explain this difference between two arrays only about 100 nm apart.

Information from Complements

A complementary pair is shown in Fig. 8a and 8b. The E-face shows both rings and trigonal depressions and there are a few particles on it. The complementary P-face shows rather irregularly shaped particles with extensive deformation, although there is some indication of craters on a few of the particles.

We have so far been unable to produce a good complementary replica pair to show the E-face structure complementary to the regularly shaped cratered P-face particles. However, from the complementary pairs that we have examined we get the impression that it is the exception rather than the rule to have regularly shaped cratered P-face particles as the complement of clear E-face rings and depressions. Complementary pairs do demonstrate that even after fracturing at $-196°C$ there is considerable plastic deformation, as has also been shown by Sleytr and his colleagues (see review by Sleytr and Robards, ref. 11). It may not be wise to assume that the regular cratered P-faces are the perfect complements of the E-face rings and depressions. Indeed, Kübler et al (7) had difficulty finding P-face structures complementary to the trigonal depressions of the E-face. They averaged the information from an E- and a P-face that were not actual complements of each other.

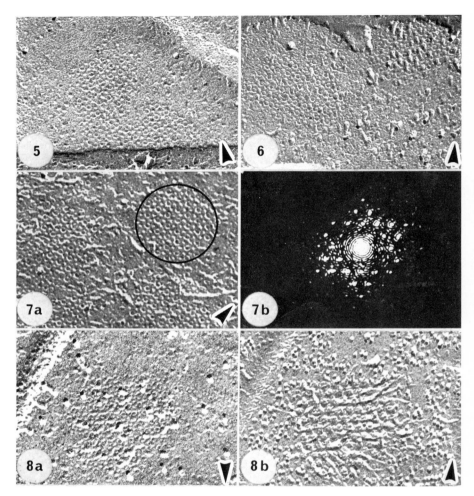

FIGS. 5–8. 5: E-face array from preparation made by suspending in distilled water and centrifuging before freezing and fracturing in a prenicked plastic tube. Each ring-like depression in the array is surrounded by small depressions at the trigonal points of the lattice. The small depressions themselves thus outline hexagons around the rings. **6:** E-face arrays from preparation made by suspending in distilled water and centrifuging before freezing and fracturing in a complementary holder. The array on the left shows rings and depressions, whereas that on the right shows some stretched particles in a regular pattern on the array. **7a:** E-face arrays from same preparation as that shown in Fig. 5. **7b:** Diffraction pattern obtained from the circled array shown in **(a).** Spots out to the fifth order are visible. **8a and 8b:** Complementary E-face **(a)** and P-face **(b)** from a preparation made by suspending in distilled water and centrifuging before freezing and fracturing. The E-face shows some particles on the array of rings and depressions. The P-face shows considerable plastic deformation of the particles, although craters can be seen on some of them. The two micrographs are direct rather than mirror images of one another. Figs. 5–8: *broad arrow,* shadowing direction; *magnification,* ×172,000.

CONCLUSIONS

Gross et al (6) obtained replicas showing regular cratered P-face particles and E-face rings and depressions after fracturing and replicating in ultrahigh vacuum. Although the cratered P-face particles may be a function of factors other than the very clean vacuum, it does seem that the preservation of the E-face structures without contamination requires a low vapor pressure of contaminants in the region of the specimen. This present work shows that a very simple and inexpensive freeze-fracturing apparatus is capable of providing adequate shielding. To be confident of the reality of structures seen on freeze-fracture faces, contamination must be eliminated with greatest care. A practical suggestion is that the contamination shielding of any freeze-fracture apparatus be tested by making yeast replicas and examining the replicas for the E-face rings and depressions. This is essentially the suggestion made some years ago by McNutt and Weinstein (8).

ACKNOWLEDGMENT

We would like to thank Dr. Uwe Sleytr for allowing us to see his manuscript (10) before publication.

REFERENCES

1. Bullivant, S. (1973): Freeze-etching and freeze-fracturing. In: *Advanced Techniques in Biological Electron Microscopy,* edited by J. K. Koehler, pp. 67–112. Springer, Berlin, Germany.
2. Bullivant, S. (1977): Evaluation of membrane structure facts and artefacts produced during freeze-fracturing. *J. Microsc.,* 111:101–116.
3. Bullivant, S., and Ames, A. (1966): A simple freeze-fracture replication method for electron microscopy. *J. Cell Biol.,* 29:435–447.
4. Chalcroft, J. P., and Bullivant, S. (1970): An interpretation of liver cell membrane and junction structure based on observation of freeze-fracture replicas of both sides of the fracture. *J. Cell Biol.,* 47:49–60.
5. Gross, H., and Moor, H. (1978): Decoration of specific sites on freeze-fractured membranes. In: *Electron Microscopy 1978,* Vol. 2, edited by J. M. Sturgess, pp. 140–141. Microscopical Society of Canada, Toronto.
6. Gross, H., Bas, E., and Moor, H. (1978): Freeze-fracturing in ultrahigh vacuum at − 196°C. *J. Cell Biol.,* 76:712–728.
7. Kübler, O., Gross, H., and Moor, H. (1978): Complementary structures of membrane fracture faces obtained by ultrahigh vacuum freeze-fracturing at − 196°C and digital image processing. *Ultramicroscopy,* 3:161–168.
8. McNutt, N. S., and Weinstein, R. S. (1971): Useful resolution standards for freeze-cleave and etch replication techniques. In: *Proc. Electron Microsc. Soc. Am.,* Vol. 29, edited by C. J. Arcenaux, pp. 444–445. Claitor's Publ. Div., Baton Rouge.
9. Mühlethaler, K., Wehrli, E., and Moor, H. (1970): Double fracturing methods for freeze-etching. In: *Microscopie électronique 1970,* Vol. 1, edited by E. L. Benedetti and P. Favard, pp. 449–450. Société Francaise de Microscopie Électronique, Paris.
10. Sleytr, U. B., and Messner, P. (1978): Freeze-fracturing in normal vacuum reveals ring-like yeast plasmalemma structures. *J. Cell Biol.,* 79:276–280.
11. Sleytr, U. B., and Robards, A. W. (1977): Plastic deformation during freeze-cleaving: A review. *J. Microsc.,* 110:1–25.

Freeze Fracture: Methods, Artifacts, and
Interpretations, edited by J. E. Rash and
C. S. Hudson. Raven Press, New York © 1979.

Low-Temperature Freeze Fracturing
to Avoid Plastic Distortion

*Stefan Kirchanski, **A. Elgsaeter, and *D. Branton

* Cell and Developmental Biology, The Biological Laboratories, Harvard University,
Cambridge, Massachusetts 02138; and ** Institute for Biophysics, University of Trondheim,
N-7034 Trondheim-NTH, Norway

After even the most rapid freezing now possible with fast freezing techniques (see R. Ornberg and T. Reese, *this volume;* D. E. Chandler, *this volume*), certain baffling inconsistencies continue to embarrass our understanding of the replica images: E-faces often do not have pits corresponding to P-face particles. In unusual membranes, such as the pellicle of *Euglena gracilis,* highly ordered striations of the E-face fail to match large, random particles on the P-face (2), and in the purple membrane of *Halobacterium halobium* the rather smooth E-faces fail to match the ordered array of P-face particles (1). An explanation for this pervasive mismatch of fracture faces is an alteration of membrane components during the fracturing process. The cleavage of frozen membranes undoubtedly releases considerable energy, that may result in local rearrangements of membrane components. Many model systems and some biological membranes show obvious plastic deformation during fracturing; Sleytr and Robards (4) have recently reviewed this subject and have shown the deformation is greatly reduced as the temperature of cleavage is reduced to that of liquid helium (approximately 4°K). We are trying to determine if low-temperature freeze fracturing can reduce plastic deformation in biological samples and improve the match between E- and P-faces. As a test sample we used the purple membranes of *Halobacterium* fractured at temperatures approaching that of liquid helium.

Figure 1 illustrates unidirectionally replicated "control" P and E faces of purple membranes fractured at approximately 150°K. Although this temperature is somewhat lower than that normally used in freeze fracture, the image of the purple membrane is as has been previously observed. The P-face shows hexagonally arranged particles with fairly common offsets or defects in the lattice. The E-face has been specially selected to show the slight detail that can be found occasionally on the normally smooth E-faces (K. Fisher, *personal communication*). In Fig. 2 we see the results of fracturing and unidirectionally replicating the membranes at 10°K. The E-face is dramatically altered, in that virtually every face (not specially selected faces, as in Fig. 1) shows considerable

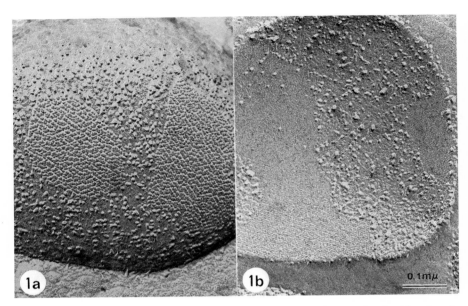

FIG. 1. Protoplasmic **(a)** and extracellular **(b)** faces of *H. halobium* fractured and unidirectionally replicated at a stage temperature of 150°K and a knife temperature of 77°K. ×118,000.

detail with long-range order apparent in several directions. Enhanced detail and order at low temperatures is also apparent in Fig. 3, which shows membranes fractured and rotary replicated at 70°K.

We have shown that freeze fracture at temperatures below that of liquid nitrogen results in the appearance of considerable detail on E-faces of purple membranes. Detail seems to improve as the temperature is lowered, thus suggesting that the smooth E-faces in conventional freeze fractures may be a result of plastic deformation and flow of membrane components. Recent work by Henderson et al. (3) demonstrated that the extracellular leaflet of the purple membrane is primarily glycolipid, whereas the protoplasmic leaflet (as has long been suspected from freeze fracture) contains all the membrane's protein. This extreme asymmetry may explain why the deformation is so obvious in this membrane. Low-temperature P-faces (Figs. 2 and 3) also appear somewhat different from the conventional P-face (Fig. 1). At this time, it is difficult to tell whether this is due to reduced plastic deformation, poor fracturing as a result of knife failure on these extremely hard samples, or changes in decoration effects as the replicating material condenses at these very low temperatures. Nevertheless, even with the improvements of ultralow-temperature fracture, the E–P-face match is still not perfect. Future experiments will reveal whether this is due to residual plastic deformation, alterations during the freezing process, or both.

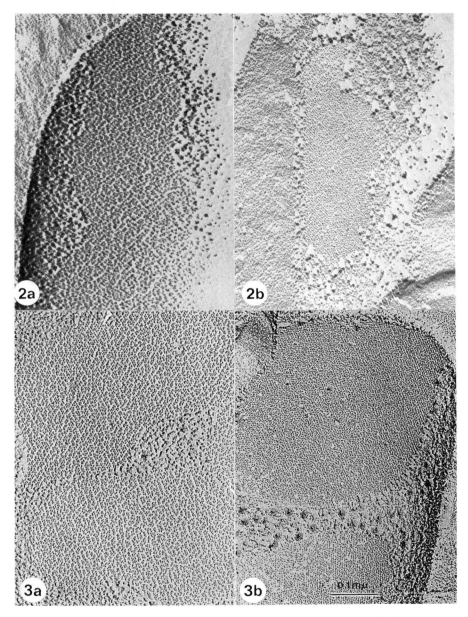

FIGS. 2 and 3. 2: Protoplasmic **(a)** and extracellular **(b)** faces of *H. halobium* fractured and unidirectionally replicated at a stage and knife temperature of 10°K. **3:** Protoplasmic **(a)** and extracellular **(b)** faces of *H. halobium* fractured and rotary replicated at a stage temperature of 70°K and a knife temperature of 10°K. Figs. 2 and 3: ×118,000.

REFERENCES

1. Blaurock, A. E., and Stoeckenius, W. (1971): Structure of the purple membrane. *Nature New Biol.,* 233:152–155.
2. Branton, D., and Kirchanski, S. (1977): Interpreting the results of freeze-etching. *J. Microsc.,* 111:117–124.
3. Henderson, R., Jubb, J. S., and Whytock, S. (1978): Specific labelling of the protein and lipid on the extracellular surface of purple membrane. *J. Mol. Biol.,* 123:259–274.
4. Sleytr, U. B., and Robards, A. W. (1977): Plastic deformation during freeze-cleavage: A review. *J. Microsc.,* 110:1–25.

*Freeze Fracture: Methods, Artifacts, and
Interpretations,* edited by J. E. Rash and
C. S. Hudson. Raven Press, New York © 1979.

The Sectioned-Replica Technique: Direct Correlation of Freeze-Fracture Replicas and Conventional Thin-Section Images

John E. Rash

*Department of Pharmacology and Experimental Therapeutics, University of Maryland
School of Medicine, Baltimore, Maryland 21201*

It is widely presumed that freeze-fracture and freeze-etch replicas may be examined only after complete removal of all adhering tissue or subcellular debris. This assumption is not entirely accurate. By combining the freeze-fracture and conventional thin-section imaging techniques, for example, it is possible to correlate directly, in a single micrograph, the unique structural and compositional information provided by the separate techniques. Previous efforts to combine the two methods (2,8,9,20) were only marginally successful, primarily because they employed *cross-sections* of the replica-tissue interface and attempted to identify only the precise location of the fracture plane. Interestingly, two of those studies concluded that the replicating material was deposited on true membrane surfaces (8,20), whereas the other two concluded that the fracture faces result from splitting of the original membrane bilayers (2,9). In contrast, this report describes a reliable method for obtaining thin sections approximately parallel to the replica-tissue interface, each section containing sufficiently large areas of replicated surface detail and subjacent cytoplasmic architecture to permit direct and unambiguous correlation of the same structure by the two independent methods. By carefully examining these "sectioned replicas" in stereo electron micrographs, well-characterized cytoplasmic and surface markers may be used to help differentiate between similar cell types or cytoplasmic structures that are otherwise indistinguishable by either technique alone. Finally, it should be noted that since cytoplasmic structures are retained within the sectioned replicas, a variety of conventional thin-section labeling techniques may now be employed to identify individual structures and macromolecules exposed at the replica-tissue interface. [Preliminary data concerning direct and indirect labeling of particles and pits have been published elsewhere (12,15,16).]

MATERIALS AND METHODS

Adult and 31-day postnatal rats were prepared by perfusion for 1 min with oxygenated rat Ringers solution (pH 7.3) containing 10 U/ml of heparin, followed immediately by perfusion fixation with either 2.5% glutaraldehyde or 2% formaldehyde in oxygenated rat Ringers solution (pH 7.3). After fixation for up to 1 hr, dissected samples containing identified neuromuscular junctions (for methods see ref. 12) were slowly infiltrated by dropwise additions of 30% glycerol in Ringers solution, equilibrated to 30% glycerol, placed on Balzers 3-mm-gold specimen supports, and subsequently frozen in Freon-12 maintained at its freezing point (− 150°C) in a bath of liquid nitrogen. (For more complete details, see Hudson et al., *this volume.*)

All samples were cleaved at − 105°C in a Balzers 360 M freeze-etch device equipped with an electron beam gun for platinum deposition, a resistance unit for carbon deposition, a quartz crystal thin-film monitor for regulating replica thickness, and an automatic shutter for synchronizing the onset of platinum deposition with the final cleaving step (Ellisman and Staehelin; and Rash et al., *this volume*). The quartz crystal monitor cutoff frequency normally used is 125 Hz. However, to resolve cytoplasmic detail beneath the very electron-dense platinum replica, the thickness of the platinum layer was reduced approximately 25% by using a cutoff frequency of 60 to 100 Hz.

After replication, surfaces were stabilized with vaporized carbon *or were left uncoated* [useful for post fracture labeling of exposed particles (14)]. The samples were thawed, floated in a bath containing 30% glycerol plus 1% osmium tetroxide, postfixed for 1 hr, and either (a) immersed, rinsed, and poststained in 0.5% aqueous unbuffered uranyl acetate, (b) dehydrated immediately in graded methanol series and poststained with 3% uranyl acetate in absolute methanol, or (c) left unstained. After dehydration and three rinses in absolute methanol, samples were transferred to acetone and through 30 and 70% plastic mixtures in acetone. The final plastic mixture consisted of 10% Epon 812, 20% Araldite 6005, 70% dodecenyl succinic anhydride (DDSA), plus 1.5% DMP-30 added as catalyst. To ensure complete penetration of the plastic into the rather impervious carbon-platinum replicas, and thereby to prevent splitting of the sections at the replica-plastic interface, unpolymerized specimens were briefly evacuated three times to 10 torr and alternately repressurized to 760 torr. The samples were polymerized for 24 hr at 70°C and 500 torr. Silver to pale gold sections were cut approximately parallel to the original replica surface using a Sorval MT-2 ultramicrotome. Sections were picked up on uncoated 200 mesh grids or on silver-gray collodion films on 1 × 2 mm single hole supports. Samples were photographed at 80 kV in a Siemens Elmiskop 101 electron microscope equipped with a ± 24 double-tilt device (goniometer).

Method for Recovery of Samples After Accidental Thawing

It is well established that cryoprotected samples refrozen immediately after accidental thawing exhibit moderate to severe cell disruption, presumably caused by ice crystals formed as a result of water-glycerol phase separation during thawing and refreezing. To demonstrate that the deleterious effects of improper *refreezing* may be reduced or eliminated, test samples of glutaraldehyde-fixed, glycerinated muscle were thawed rapidly, reequilibrated for 30 min with fresh 30% glycerol in rat Ringers solution, refrozen in a slurry of freezing Freon-12, and fractured and replicated according to conventional procedures.

RESULTS AND DISCUSSION

At very low magnification, thin sections cut approximately parallel to and containing portions of the tissue-replica interface revealed discontinuous muscle fibers surrounded by thin, seemingly worthless "wisps" of replica (Fig. 1a). At magnifications sufficient to reveal intramembranous particles (IMPs), however, the sectioned replicas occupied a significant portion of the micrograph area (Fig. 1b). Despite some microwrinkling of replicated surfaces, preservation of both replica and tissues was considered adequate for recognizing specific classes of IMPs and their respective pits. For example, several "square arrays" (13) of 60-Å P-face particles (Fig. 1c) were discerned in the sarcolemma of a fast-twitch fiber from a 31-day postnatal rat.

Because a large number of thin sections are obtained from the replica-tissue interface, the sectioned-replica technique may be utilized to greatest advantage when serial sections are analyzed. The large number of sections obtained from the replica-tissue interface greatly increases the probability of obtaining images from very small or very rare structures. For example, many serial sections were obtained from a minute neuromuscular synapse of a 31-day postnatal rat (Fig. 2a). Despite brief fixation in formaldehyde, glycerination, freezing, and thawing, the quality of tissue preservation was comparable to that obtained by many conventional techniques. In higher magnification stereo pairs (Fig. 2b), the replicated synaptic vesicle membrane E- and P-faces were seen to be continuous with their conventional thin-section images in the underlying cytoplasm. At still higher magnification (Fig. 2c), a layer equivalent to a half-membrane thickness was seen to be continuous with the replicated surface of a cytoplasmic vesicle. (Note the much thicker mitochondrial membrane: large arrowhead.) Thus, it is clear that by carefully examining sectioned replicas, it is possible to discriminate between structures that heretofore were indistinguishable in conventional freeze-fracture replicas. For example, replicated synaptic vesicles (both free and fusing with the surface membrane), coated vesicles, and coated pits have been discerned in high-magnification stereopairs. From these and similar preparations, we are currently attempting to identify specific IMP profiles that may be of value in discriminating between these two types of vesicles in conventional freeze-fracture replicas (17). (For a description of synaptic vesicle membrane recycling via a coated vesicle mechanism, see refs. 3 and 4.)

Accidentally Thawed Specimens Need Not Be Discarded

To demonstrate the quality of tissue preservation attainable after two carefully controlled freeze-thaw cycles, a sample of glutaraldehyde-fixed, glycerinated

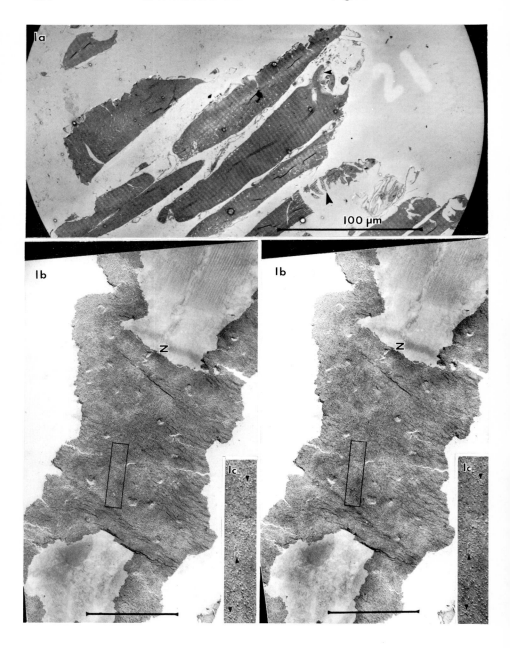

muscle was thawed, reequilibrated to 30% glycerol, refrozen, fractured, and replicated. A node of Ranvier embedded approximately 100 μm below the replicated surface (Fig. 3) revealed preservation of cellular detail equivalent to that in the best preparations obtained by conventional fixation, staining, and embedding techniques. Clearly, glutaraldehyde-fixed, glycerol-infiltrated tissues examined in thin sections do not exhibit excessive damage following carefully controlled freezing and thawing. (Analysis of conventional freeze-fracture replicas prepared after two freeze-thaw cycles is not yet completed, but preliminary data suggest that many small areas of membrane disruption develop despite the most stringent precautions. Thus, it is concluded that accidentally or partially thawed freeze-fracture samples may be recovered for conventional thin-section analysis, but may not be suitable for detailed freeze-fracture studies.)

CONCLUSIONS

The sectioned-replica technique is demonstrated and a method for direct correlation of freeze-fracture and conventional thin-section images is described. If IMPs represent protein or lipoprotein complexes (1,5,10), and if significant biochemical activity or immunological specificity remain after weak aldehyde fixation (18,19), the nonreplicated (unshadowed) portions of some intramembranous particles may be sufficiently exposed by fracturing to permit direct labeling by immunological, pharmacological, or autoradiographic methods. Preliminary data from prefracture (11,12,15) and postfracture labeling experiments (14,16) suggest the efficacy of such approaches.

ACKNOWLEDGMENTS

This work was supported by grants from the National Institutes of Health no. NS#14648, and by the Muscular Dystrophy Association.

FIG. 1. a: Very low-magnification electron micrograph of a thin section from the replica-tissue interface. Thin "wisps" of replica *(large arrowhead)* are difficult to resolve at this magnification. The etching power of the electron microscope beam at condensor cross-over was used to mark individual motor endplates *(small arrowhead)* for subsequent serial section reconstruction. ×480. **b:** Stereoelectron micrograph of a sectioned replica revealing various classes of IMPs in the sarcolemma of a 31-day postnatal rat extensor digitorum longus muscle. The appearance of "square arrays" in freeze-etch replicas of developing muscle fibers (6,7) may now be correlated directly with the development of characteristic cytoplasmic features, such as Z-band width. (Note the immature myofibril extending beneath the replica in this unstained section.) The extent of specimen compression during thin sectioning is reflected in "microwrinkling" or the replicated sarcolemma. ×30,000. **c:** Higher magnification of P-face square arrays *(arrowheads)*. Although there is a slight loss of resolution in sectioned replicas, individual particles and pits may be discerned in all fracture faces. (**c** is a photographic enlargement of **b.**) ×75,000.

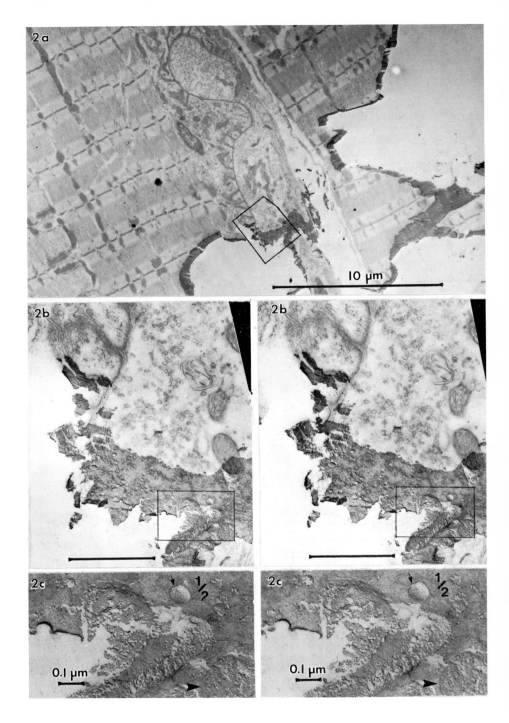

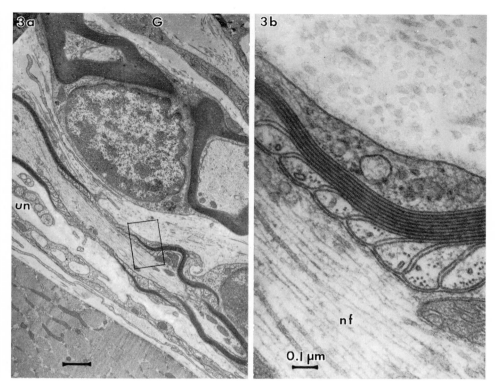

FIG. 3. a and b: Thin section from a specimen that was thawed, reequilibrated with 30% glycerol, and refrozen. Note the absence of detectable membrane disruption. (Partial reannealing of membrane microdisruptions prior to osmium tetroxide postfixation is presumed.) un, unmyelinated nerves; G, Golgi apparatus; nf, neurofilaments. **a:** ×7,600; **b:** ×70,000.

REFERENCES

1. Branton, D. (1971): Freeze-etching studies of membrane structure. *Philos. Trans. Soc. Lond.* [*Biol.*], 261:133–138.
2. Bullivant, S. (1973): Freeze-etching and freeze-fracturing. In: *Advanced Techniques in Biological Electron Microscopy,* edited by J. K. Koehler, pp. 67–112. Springer, Berlin.
3, Heuser, J. E., and Reese, T. S. (1973): Evidence for recycling of synaptic vesicle membrane during transmitter release at the frog neuromuscular junction. *J. Cell Biol.,* 57:315–344.

FIG. 2. a: Sectioned replica obtained from the neuromuscular junction of a formaldehyde-fixed 31-day postnatal rat. ×5,000. **2b:** In higher magnification stereomicrographs, surface replicas of synaptic vesicles and mitochondria are continuous with their corresponding conventional thin-section images, thereby allowing simultaneous and unambiguous correlation of biological structures by two independent imaging techniques. ×25,000. **2c:** High-magnification stereomicrographs reveal the direct continuity of a replicated membrane face with the electron lucent layer in the subjacent thin section. Only the outer half of the membrane *(small arrow)* is resolved in the thin section. ×72,500.

4. Heuser, J. E., Reese, T. S., and Landis, M. D. (1974): Functional changes in frog neuromuscular junctions studied with freeze-fracture. *J. Neurocytol.,* 3:108–131.
5. Hong, K., and Hubbell, W. L. (1972): Preparation and properties of phospholipid bilayers containing rhodopsin. *Proc. Natl. Acad. Sci. USA,* 69:2617–2621.
6. Hudson, C. S., Dyas, B. K., and Rash, J. E. (1978): Changes in the number and distribution of square arrays during postnatal rat fiber type differentiation. *J. Cell Biol.,* 79:37a.
7. Hudson, C. S., Rash, J. E., and Albuquerque, E. X. (1977): A thin section and freeze-fracture study of mammalian neuromuscular junction development. *J. Cell Biol.* 75:116a.
8. Leak, L. V. (1969): Path of fracture planes along membrane surfaces in frozen-etched tissue. *EMSA Proceedings,* 27:334–335.
9. Nanninga, N. (1971): Uniqueness and location of the fracture plane in the plasma membrane of *Bacillus subtilis. J. Cell Biol.,* 49:564–570.
10. Pinto Da Silva, P., Douglas, S. D., and Branton, D. (1971): Localization of A antigen sites on human erythrocyte ghosts. *Nature,* 232:194–195.
11. Rash, J. E., Copio, D., and Eldefrawi, M. E. (1979): Freeze-fracture autoradiography by the sectioned-replica technique. *J. Cell Biol. (in preparation).*
12. Rash, J. E., Copio, D. S., Eldefrawi, M. E., and Hudson, C. S. (1977): Practical labelling techniques for freeze-fracture. *J. Cell Biol.,* 75:247a.
13. Rash, J. E., and Ellisman, M. H. (1974): Studies of excitable membranes. *J. Cell Biol.,* 63:567–586.
14. Rash, J. E., Hudson, C. S., Eldefrawi, M. E., and Graham, W. F. (1979): Post-fracture labelling of intramembranous particles by the sectioned-replica technique. *J. Cell Biol. (in preparation).*
15. Rash, J. E., Hudson, C. S., and Ellisman, M. H. (1978): Ultrastructure of acetylcholine receptors at the mammalian neuromuscular junction. In: *Cell Membrane Receptors for Drugs and Hormones: A Multidisciplinary Approach,* edited by R. W. Straub and L. Bolis, pp. 47–68. Raven Press, New York.
16. Rash, J. E., Hudson, C. S., and Graham, W. F. (1978): Direct and indirect labelling of particles and pits in freeze-etch replicas. *J. Cell Biol.,* 79:232a.
17. Rash, J. E., Warnick, J. E., Albuquerque, E. X., and Ellisman, M. H. (1976): Freeze-fracture studies of quiescent, stimulated, briefly rested, and toxin-activated rat neuromuscular junctions. *J. Cell Biol.,* 70:303a.
18. Sabatini, D. D., Bensch, K., and Barrnett, R. J. (1963): Cytochemistry and electron microscopy: The preservation of cellular ultrastructure and enzymatic activity by aldehyde fixation. *J. Cell Biol.,* 17:19–58.
19. Sternberger, L. A. (1967): Electron microscopic immunocytochemistry: A review. *J. Histochem. Cytochem.* 15:139–159.
20. Weinstein, R. S. (1969): Electron microscopy of surfaces of red cell membranes. In: *Red Cell Membrane: Structure and Function,* edited by G. A. Jamieson and T. J. Greenwalt, pp. 36–76. J. B. Lippincott, Philadelphia.

Freeze Fracture: Methods, Artifacts, and Interpretations, edited by J. E. Rash and C. S. Hudson. Raven Press, New York © 1979.

Use of Double-Tilt Device (Goniometer) to Obtain Optimum Contrast in Freeze-Fracture Replicas

*Russell L. Steere and **John E. Rash

*Plant Virology Laboratory, Plant Protection Institute, AR, SEA, U.S. Department of Agriculture, Beltsville, Maryland 20705; and **Department of Pharmacology and Experimental Therapeutics, University of Maryland School of Medicine, Baltimore, Maryland 21201*

The images of transmission electron microscopes are formed by exposure of the fluorescent screen or the photographic emulsion to the electrons that manage to penetrate the specimen without being scattered to the walls or blocked by the apertures. The images of freeze-fracture or freeze-etch replicas result from local differences in scattering of electrons by the thin "shadowing" layer of heavy metal (1,5). Platinum is generally used for high-resolution replicas, although chromium was initially employed (3). If the layer were applied uniformly from a source directly above (i.e., perpendicular or normal to the exposed surface), the incident electron beam would traverse a layer of nearly uniform relative thickness (Fig. 1, top) and thus the resulting images would have little or no contrast (Fig. 1, bottom). (At the edges of particles, a slight "snowdrift" effect may occur, possibly resulting in a slight increase in edge contrast.) To provide the desired contrast enhancement, the platinum shadow layer is usually deposited at an angle of 45° (Fig. 2), thereby producing a layer of varying relative thickness

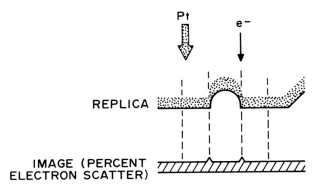

FIG. 1. Unidirectional deposition of a platinum replica *(top)* on an irregular plane oriented perpendicular to the direction of shadow. The "image" *(bottom)* is a diagrammatic representation of the relative amount of electron scatter at each point along the replica.

161

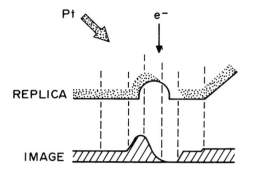

FIG. 2. Platinum replica deposited at a 45° angle to the plane of the replica *(top)* and viewed with the electron beam normal to the replica *(bottom)*.

when viewed with the electron beam perpendicular to the shadowed surface (approximately 45° from the shadowing source).

Fracture faces are often very irregular as a result of deviations in the fracture plane. In addition, the replicas may deform or collapse after removal of the supporting biological material or when the cleaned replicas are dried onto the grids. Thus, even when the electron beam is perpendicular to the grid, there are often areas of the replica through which the passage of the electron beam is far from perpendicular or differs significantly from 45° to the original shadowing angle. It is appropriate, therefore, to compare the effects on apparent specimen contrast and detail when a specimen shadowed at 45° is examined with the electron beam at different angles: path "a," 45° to the shadowing angle (Fig. 3a); path "b," nearly perpendicular to the original shadowing angle (Fig. 3b); or path "c," parallel to the shadowing angle (Fig. 3c).

Electrons traversing path "a" (Fig. 3a) would encounter variable numbers of platinum grains and the resulting micrographs would have continuous tone and the complete range of photographic contrasts. Electrons traversing path "b" (perpendicular to the original shadowing angle, Fig. 3b) would encounter few platinum grains while traversing the unshadowed base of the stylized particle, but would be entirely blocked (inelastic and elastic scatter) by the numerous superimposed platinum grains on the heavily shadowed crests of the particles. Such images have excessively high contrast and discontinuous tone. Finally, electrons traversing path "c" (parallel to the original shadowing angle, Fig. 3c), would have a nearly equal chance of encountering platinum atoms. The resulting images have little visual or photographic contrast. The optimum *local viewing angle,* therefore, is about 45° to the local shadowing angle (path "a" or half way between "b" and "c." Contrast for each portion of a specimen may be optimized by appropriate tilting of the specimen by using a goniometer or double-tilting stage.

In practice, replicas with considerable contour often have large areas that are tilted to such a degree that the entire area (not just individual particles) scatter most of the electrons and reveal images with little or no fine structural detail (see Fig. 18.7 in ref. 4). In some instances, the part of a replica from

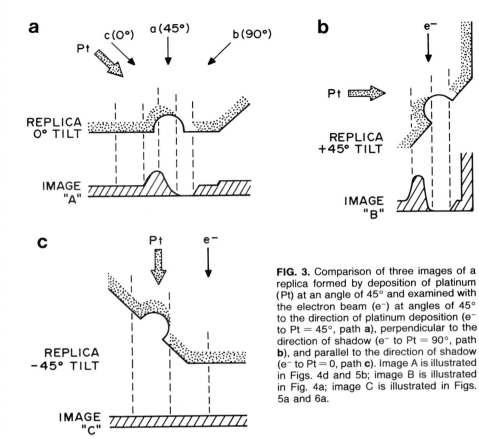

FIG. 3. Comparison of three images of a replica formed by deposition of platinum (Pt) at an angle of 45° and examined with the electron beam (e⁻) at angles of 45° to the direction of platinum deposition (e⁻ to Pt = 45°, path **a**), perpendicular to the direction of shadow (e⁻ to Pt = 90°, path **b**), and parallel to the direction of shadow (e⁻ to Pt = 0, path **c**). Image A is illustrated in Figs. 4d and 5b; image B is illustrated in Fig. 4a; image C is illustrated in Figs. 5a and 6a.

which useful information is desired is tilted to such a degree that the replica actually appears to be worthless. Fortunately, the deep tilt or goniometer stages in most new transmission electron microscopes (TEMs) permit tilting the specimen within the microscope while observing the image on a fluorescent screen. With such a device it is convenient to obtain stereoimages that demonstrate clearly the changes in appearance of specimens as they are tilted in different directions. This is clearly demonstrated in Figs. 4a to 4d in which the tilt between prints of each stereopair is approximately 10° and in which the tilt between successive stereopairs is approximately 25°.

The objects of Fig. 4a are sufficiently light and the shadows sufficiently dark that one might easily reject such a specimen thinking that the pile-up of platinum in the white areas (shadows being black) is so great that it obliterates all fine structural details. Tilting the entire specimen approximately 25° in the proper direction (Fig. 4b) partially corrects the misconception but still reveals less than the desired amount of information. Tilting the specimen another 25° in

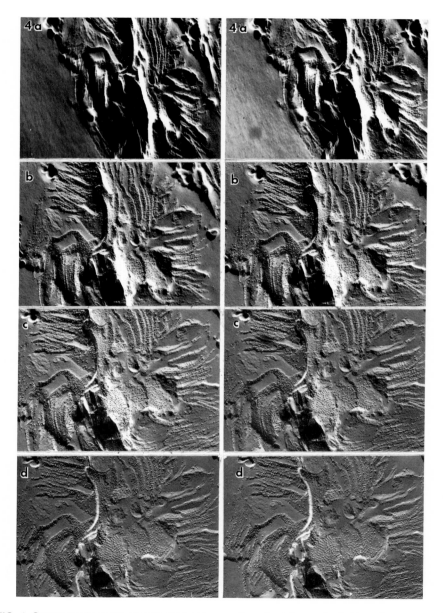

FIG. 4. Stereoscopic views of a freeze-fracture replica of a tobacco leaf chloroplast mounted on a grid tilted at a goniometer reading of approximately $+50°$ (**4a**), $+25°$ **(b)**, $0°$ **(c)**, and $-25°$ **(d)**. (Goniometer tilt angles are measured with respect to a plane perpendicular to the electron beam axis.) Thus, optimum contrast in this region of the replica is obtained at a goniometer tilt of $-25°$, (equivalent to image A, Fig. 3); JEM 100B, $±45°$ top-entry goniometer; black shadows. ×45,000

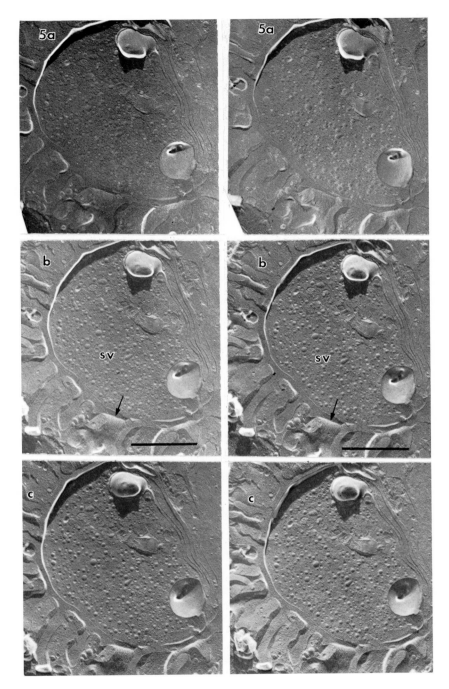

FIG. 5. Cross-fractured neuromuscular junction from rat skeletal muscle, black shadows (See Rash et al., 1978.) The replica was photographed with the grid tilted at goniometer readings of approximately $-30°$ **(a)**, $-5°$ **(b)**, and $+20°$ **(c)**. (Siemens 101, $\pm 24°$ double-tilt device.) Note that optimum contrast is obtained at a goniometer tilt of $-5°$ and an e^- to Pt angle of $45°$ **(b)**. Junctional folds *(arrows)* and synaptic vesicles (SV). $\times 17,000$.

the same direction (Fig. 4c) (the horizontal position in the goniometer) provides even more improvement in the image. Finally, tilting the specimen approximately 25° further in the same direction (Fig. 4d) assures us that the replica was not heavily overshadowed, but is actually a high-quality replica. This sequence illustrates electron-beam angles "a" (Fig. 4d) through "b" (Fig. 4a) and documents

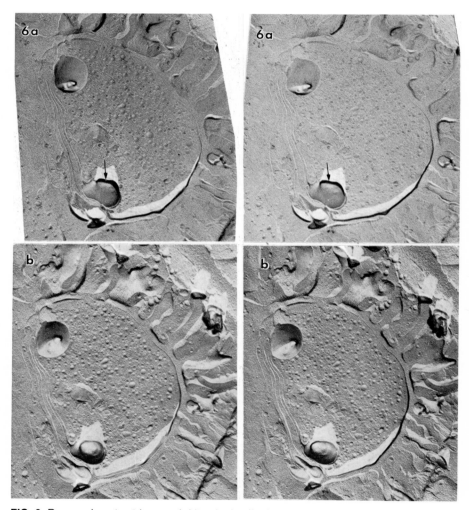

FIG. 6. Reversed contrast images (white shadows) of the areas depicted in Figs. 5a and 5c. The irregularly shaped stereoimages **(a)** result from the significantly different views obtained at steep angles of tilt. The thick carbon coat *(arrow)* resulted from prolonged exposure to the electron microscope beam without proper protection by a liquid nitrogen cold trap. ×17,000. (Fig. 6b modified from Rash, et al., ref. 2.)

excessive specimen contrast at viewing angles greater than 45° to the shadowing angle.

In other situations, the specimen has a washed-out appearance when first viewed in the electron microscope (Fig. 5a) and, without proper goniometric analysis, might be discarded as worthless. When such specimens are tilted far enough in the proper direction, however, they often reveal the crispness of high-quality replicas (Fig. 5b). Further tilting (Fig. 5c) produces excessive contrast. Thus, this sequence documents reduced specimen contrast and reduced ultrastructural detail at electron beam angles approximately 20° less than 45° to the local shadowing angle and documents excessive specimen contrast at electron beam angles approximately 20° greater than 45° to the local shadowing angle. The apparent loss of contrast and detail is equally obvious when the images are presented with white "shadows" (Fig. 6). In such cases, the stereoimaging technique applied to images with black shadows may be helpful in the interpretation of complex three-dimensional structures and resolving controversies concerning the shapes of small structures partially obscured in "shadow" (i.e., not replicated with platinum).

One last note of caution should be given: Because of the great three-dimensional complexity of replicas, significant misinterpretation of structural detail is possible in a single pair of stereoelectron micrographs. Consequently, multiple-viewing angles may be required for proper interpretation (see especially Fig. 18.16 in ref. 4). Regardless of the use of the stereoviewing techniques, however, goniometric analysis may be required to obtain the optimum electron beam angle for maximum retrieval of pertinent biological information.

REFERENCES

1. Müller, H. O. (1942): Die Ausmessung der Tiefe übermikroskopischer Objekte. *Kolloid-Z.*, 99:6–28.
2. Rash, J. E., Hudson, C. S. and Ellisman, M. H. (1978): Ultrastructure of acetylcholine receptors at the mammalian neuromuscular junction. *In: Cell Membrane Receptors for Drugs and Hormones: A Multidisciplinary Approach,* edited by R. W. Straub and L. Bolis, pp. 47–68. Raven Press, New York.
3. Steere, R. L. (1957): Electron microscopy of structural detail in frozen biological specimens. *J. Biophys. Biochem. Cytol.* 3:45–60.
4. Steere, R. L. (1973): Preparation of high-resolution freeze-etch, freeze-fracture, frozen-surface, and freeze-dried replicas in a single freeze-etch module, and the use of stereo electron microscopy to obtain maximum information from them. In: *Freeze-Etching Techniques and Applications,* Chapter 18, edited by E. L. Benedetti and P. Favard, pp. 223–255. Société Française de Microscopie Electronique, Paris,
5. Williams, R. C., and Wyckoff, R. W. G. (1944): The thickness of electron microscopic objects. *J. Appl. Phys.,* 15:712–715.

Freeze Fracture: Methods, Artifacts, and Interpretations, edited by J. E. Rash and C. S. Hudson. Raven Press, New York © 1979.

Contribution of Carbon to the Image in Freeze-Fracture Replication

Joel B. Sheffield

Department of Biology, Temple University, Philadelphia, Pennsylvania 19122

The replication phase of the freeze fracture replication involves two steps: (a) the deposit of a shadow of platinum–carbon from an angle of 45° onto the exposed fracture face, and (b) the evaporation of a support film of carbon. With some freeze-fracture apparatus it is possible to move the specimen during the carbon evaporation phase so as to allow the carbon to reach areas of the sample that might otherwise be obscured by overlapping structures. It is generally assumed that the carbon contributes little to the image, although recent reviews by Stolinski (3,4) describe the visualization of carbon in areas of minimal platinum deposit. As part of a series of freeze-fracture and cell-surface replica studies, we have observed that, at times, the carbon film becomes a major source of information in the image, even in well-shadowed preparations. We have further investigated this phenomenon by using latex beads as a model system. We have compared the appearance of the beads either untreated or coated with carbon films. In some cases, latex was removed from the replicas by extraction with acetone.

MATERIALS AND METHODS

All evaporative procedures were carried out in a Denton Vacuum evaporator fitted with the Denton Freeze-Etch Module. The evaporations were carried out at a vacuum of 10^{-6} torr or better. Latex beads, with an average diameter of 850 Å, were suspended in distilled water and applied either to a carbon-coated grid or to a cleaned glass slide, approximately 1 cm². After drying, the grids were observed in the electron microscope with no further treatment. The glass slide was mounted in the freeze-etch apparatus and films of carbon of various thicknesses applied from a distance of 6 cm. The relative thickness of the film was determined by measuring the optical density at 550 Å. The films were then floated off the glass and picked up on uncoated grids. Some of these grids were then passed through several changes of acetone to dissolve the latex beads. Samples were observed in a Phillips 300 electron microscope at 60 kV accelerating voltage and images were recorded on Kodak electron image film. Replicas were also prepared of cells of the mouse mammary tumor cell line BALB/cfC3H after fixation and critical point drying and of RIII milk MuMTV after freeze etching (2).

RESULTS

There were two major effects of the carbon coat, and both were seen, to different degrees, in all samples. In comparison with the uncoated beads (Fig. 1a), the carbon-coated beads were of larger diameter and appeared to have a ring around the bead, forming a "halo" around the particle. This can be seen clearly in Fig. 1b. The width of the ring was a function of the thickness of the carbon coat and averaged 110 Å. In other areas of the film, a distinct shadowing effect could be seen (Fig. 1c). The angle of shadowing was relatively steep, corresponding to the displacement of the sample (up to 0.5 cm) from the point of true normal incidence. The assymetric deposit of carbon could be seen even more clearly after the latex beads were removed from the replica (Fig. 1d).

Figure 2 presents examples of two situations in which the effect of carbon is pronounced in carbon–platinum replicas. Figure 2a contains a replica of the mouse mammary tumor virus prepared by freeze etching in the presence of dimethyl sulfoxide (2). The areas of contour indicated by the arrows represent the effect of carbon in the image and appear as relatively dense lines at the periphery of the particle in the region away from the shadowing source. Figure 2b is a reversed contrast transmission electron micrograph of a carbon–platinum replica of a critical-point dried cell from the Balb/cfC3H line. In preparing the replica, the sample was tilted after evaporation of platinum, and rotated while the carbon was evaporated. Although there is a clear shadowing effect in several regions, the microvilli on the surface of the cell are not shadowed, but show heightened contrast at all edges. The appearance of this image is similar to that of a scanning electron micrograph, in which contour shifts are responsible for changes in contrast.

DISCUSSION

Although generally overlooked, the contribution of the carbon support film to the image in freeze fracture or conventional replication may be substantial. *A priori,* this is somewhat surprising in view of the low atomic number of carbon, 6, versus platinum, 78. In Bradley's review of shadowing procedures (1) the relative contrast per unit thickness of carbon relative to platinum is listed as 0.5:6.3. However, in preparing a replica of sufficient strength to withstand the processes of cleaning, a platinum coat of the order of 10 to 50 Å, and a carbon film of 200-Å thickness is typically applied. When an object is of the dimensions of 1,000 Å, the thickness of the carbon in regions of contour change can cause as significant an amount of contrast as the shadowing material itself.

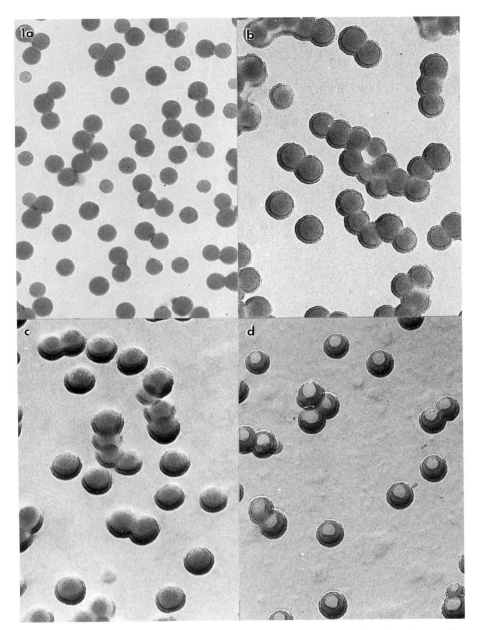

FIG. 1. Latex beads, **a** viewed directly after deposit onto a carbon film; **b** and **c** after coating with approximately 130 Å of carbon; **d** after coating with approximately 180 Å of carbon and extracting with acetone. ×84,000.

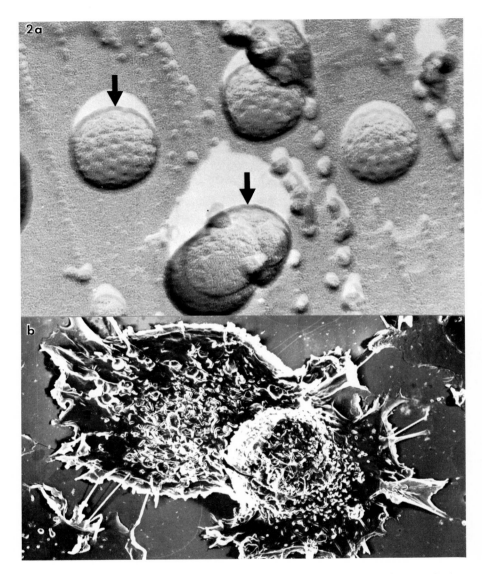

FIG. 2. a: Freeze-etch replica of milk-derived RIII MuMTV. The arrows indicate the contribution of the carbon film to the image. ×163,000. **b:** reversed contrast image of a replica of a critical point dried cell from the BACB/CFC3H line. ×5,000.

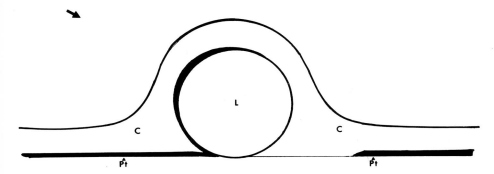

FIG. 3. Scale representation of a side view of a replica of a spherical particle, 1,000 Å in diameter with 25 Å of platinum as a shadow, and 200 Å of carbon as a support film. The shadow source is at the upper left. Platinum is represented as dense black and carbon is unshaded.

This is presented diagramatically in Fig. 3, in which an object of a diameter of 1,000 Å is shadowed with 25 Å of platinum and then coated with a uniform coat of 200 Å of carbon. On the side of the particle away from the shadowing source, one would view through a thickness of almost 1,000 Å of carbon, and this would lead to a relative level of contrast of $1,000 \times 0.5 = 500$ for carbon compared to $25 \times 6.3 = 157.3$ from 25 Å of platinum. Thus the carbon, in this region, would contribute approximately three and a half times more contrast than the platinum. The effect of this would be to create an artificial dense line at the edge of the particle, just as was seen in the replicas of virus. This effect would extend around the entire periphery of the particle, appearing as a ring unless other sources of contrast, such as a platinum shadow are present. For samples much smaller than 1,000 Å, this effect would be less pronounced. However, viruses are frequently this size, as are many cellular projections such as microvilli and retraction fibrils. For samples such as these the effect of the carbon film must be taken into account during analysis of the images.

REFERENCES

1. Bradley, D. E. (1965): Replica and shadowing techniques. In: *Techniques for Electron Microscopy,* 2nd ed., edited by D. H. Kay pp. 96–152. F. A. Davis, Philadelphia, Pa.
2. Sheffield, J. B. (1973): Envelope of mouse mammary tumor virus studied by freeze-etching and freeze-fracture techniques. *J. Virol.,* 12:616–624.
3. Stolinsky, C. (1970): Freeze fracture replication in biological research: Development, current practice, and future prospects. *Micron,* 8:87–111.
4. Stolinsky, C., and Breathnach, A. S. (1975): *Freeze-Fracture Replication of Biological Tissues. Techniques, Interpretation and Applications.* Academic Press, London.

Freeze Fracture: Methods, Artifacts, and Interpretations, edited by J. E. Rash and C. S. Hudson. Raven Press, New York © 1979.

Use of Computers in the Analysis of Intramembrane Particles

Ronald S. Weinstein, David J. Benefiel, and Bendicht U. Pauli

Department of Pathology, Rush Medical College and Rush-Presbyterian–St. Luke's Medical Center, Chicago, Illinois 60612

With the introduction of high resolution computer-controlled optical scanners, it has become feasible to build an automated system for the quantitative analysis of the intramembrane particles (IMP) that are visualized by freeze-fracture electron microscopy. We have been working on elements of such a system for several years. In order to develop the logic and software for the system, strict criteria for the identification of IMP must be established. A first step has been creation of a reference library of computer-generated synthetic electron optical images of prototypic IMP under an assortment of shadowing conditions. These images are being used to: (a) examine the effects of variables (i.e., replica thickness, IMP shape, etc.) on the electron optical appearances of IMP; (b) account for unexplained or anomalous images that are occasionally observed at real fracture faces; and (c) to assess the contributions of artifacts that may be introduced during specimen preparation and freeze-fracturing (1,7,8,16).

In this report, the methodology that we have employed to generate synthetic images of IMP for our computer reference library is summarized. Since many integral membrane proteins, visualized as IMP in freeze-fracture replicas (3,4, 14,15,21), may conform to the shape of a hemispheroid, synthetic images of such particles are described (2). Using this shape as a prototype, we have systematically examined the relationship of replica thickness to the electron optical image of a particle. Some effects of IMP clustering on particle images also are examined.

PROTOTYPIC PARTICLES

The hemispheroid was chosen as a prototype for the IMP. It is mathematically described by the formula

$$\frac{(x - x_0)^2}{A^2} + \frac{(y - y_0)^2}{B^2} + \frac{(z - z_0)^2}{C^2} = 1 \qquad [1]$$

Whereby the x, y, and z axes of the coordinate system are parallel to the A, B, and C axes of the spheroid, respectively. x_0, y_0, and z_0 are the coordinates of the spheroid center, and $z > 0$ (Fig. 1).

If the fracture face is taken as the x-y plane and the spheroid center is located at the origin, then Z representing the height of any point on the particle surface above the fracture face can be expressed as

$$Z(x,y) = C\left[1 - \left(\frac{x^2}{A^2} + \frac{y^2}{B^2}\right)\right]^{1/2} \qquad [2]$$

Algorithm

It is assumed for purposes of these simulations that the shadowing flux parallel to the x-z plane is uniform and that replicating atoms stick at their initial point of impact. The replica deposits retrograde toward the source to a thickness "I," dictated by flux and time. "I" can be resolved into displacement of the new surface toward the source in the x-direction, "I_x," and replica thickness perpendicular to the fracture face in the z-direction, "I_z." $I_z/I_x = \tan\theta$, where θ is the shadowing angle (Fig. 1).

The thickness of the replica in the z-direction is denoted $T(x,y)$ and can be calculated from $Z(x,y)$, the height above the plane of fracturing. Thus $Z(x,y)$ is zero everywhere except at a particle, where it is given by Eq. [2]. The algorithm is designed to calculate replica thickness regardless of fracture face contour, as long as it can be represented as $Z = f(x,y)$. Therefore, the algorithm can be used to simulate images of particles of any shape and, in addition, of particles in tandem (i.e., in clusters).

Modeling is done as a series of parallel cuts perpendicular to the fracture face (x-y plane), and with the axis of shadowing parallel to the x-z plane. Since it is assumed that replicating materials stick at their points of impact, each cross-section can be modeled separately and independently. For the simulations used to illustrate this paper, shadowing is at a 45° angle.

Snythetic Imaging

After generating the thickness matrix, $T(x,y)$, for the replica, a synthetic image is generated by computer. Replica thickness for each point in the matrix is translated into a gray level, using programs that take into account principles of electron optics (23) and the photographic printing process. Hard copy of the image is produced with a line printer using different characters to represent specific grey levels. Each character is spatially related to its position on the simulated fracture face.

For purposes of illustration in this paper, four gray levels were determined to be adequate to represent the major features of IMP images. Gray level designations were as follows: $T(x,y) > I_z$, 70% black; $T(x,y) \simeq I_z$, 40% black; $I_z > T(x,y) > 0$, 20% black; and $T(x,y) = 0$, white. Regions corresponding to the assigned gray levels were outlined on the $T(x,y)$ printout to define zones at each gray level. Cutouts of acetate sheets, representing the approximate gray

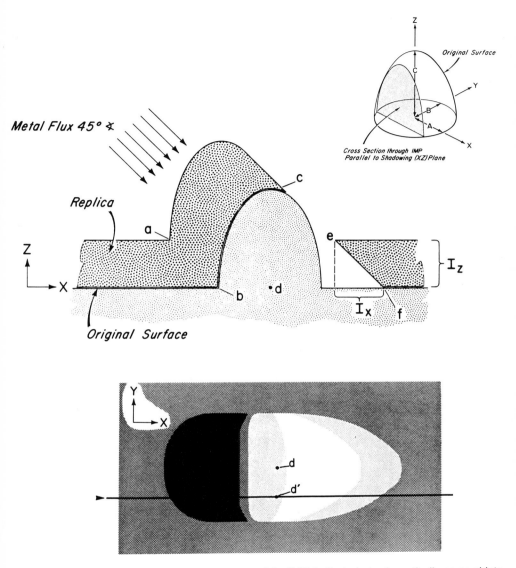

FIG. 1. *Upper right:* A prototypic intramembrane particle (IMP) is illustrated schematically as an oblate hemispheroid in isometric projection. The plane of an eccentric cross section is illustrated. *Center:* Axial eccentric cross-section of the IMP in the *x-y* plane. "c" is the point of tangency of the incident beam; "e" to "f" is the edge of the shadow; "d'" is a projection of the true geometric center "d" of the IMP. *Bottom:* Simulated image of the IMP in the *x-y* plane. The drawing is based on a synthetic "electron optical image" (at four gray levels) of the IMP above. "d" is the geometric center of the IMP and "d'" is a projection of "d" in the plane of the section. Simulation conditions: $I_x = I_z = 0.84B$. Shadowing in all simulations is at an angle of 45°.

levels and duplicating the shapes of the zones in the printout, were assembled on white sheets to produce the final simulated images.

RESULTS

The synthetic images described herein serve to illustrate basic features of the electron microscopic image of the quadratic hemispheroid, as demonstrated by unidirectional metal shadowing, and to underline some characteristics of freeze-fracture electron optical imaging. In Fig. 2, synthetic images of particles shaped as quadratic hemispheroids along with their corresponding axial cross-sections are shown. The principle components of each image are: the metal

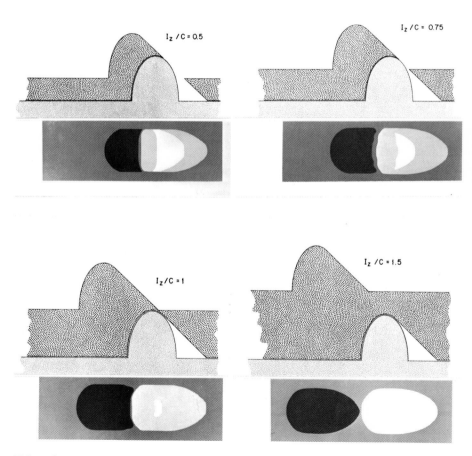

FIG. 2. Synthetic computer-generated "electron optical images" of an IMP, with incremental increases in replica thickness. Replica thicknesses (I_z) range from 0.5 to 1.5 times the IMP height (C). Axial cross-sections that hemisect the IMPs are illustrated for purposes of comparison.

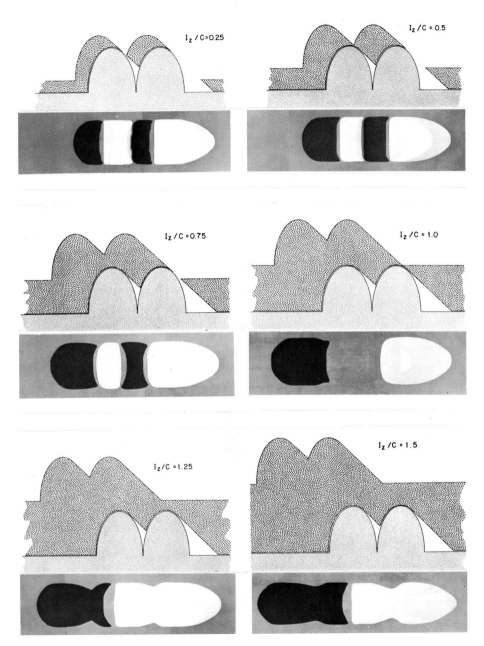

FIG. 3. Axial cross-sections and synthetic images of IMPs in tandem, with incremental increases in replica thickness. Replica thickness profoundly influences the IMP images.

cap, the relative shadow, the absolute shadow, and the background. "Relative shadow" refers to the entire zone in which platinum accumulation is diminished. The "absolute shadow" is that part of the relative shadow that is totally devoid of platinum ($I_z = 0$) and thus appears white.

The effects of incremental increases in replica thickness on the particle's image are demonstrated. Major features are as follows: (a) there is an increase in metal cap length with increased replica thickness; (b) the relative shadow remains unchanged in area and shape but the absolute shadow decreases progressively in length; (d) where replica thickness significantly exceeds particle height $(I_z > C)$, the metal cap and the shadow region appear to dissociate; and (e) at all replica thicknesses, particle width is unaltered.

In another set of simulations (not illustrated), the influence of particle height on the projected image was tested. Shadow length increases with an increase in particle height but there is no increase in metal cap length.

Fig. 3 illustrates the simulated images of two particles in tandem as is frequently encountered in IMP clusters in biological membranes. Through simulation it is shown that with relatively thin replicas the shadow of the particle proximal to the Pt–C source is shortened. Likewise, the cap region of the distal particle is shortened. As replica thickness increases, the cohort effect becomes more pronounced. The metal cap of the proximal particle increases in length but the distal cap does not. At even greater replica thicknesses, the two caps merge forming a single elongated cap and the shadow regions fuse to form a single shadow. There is a total separation of the cap and shadow regions, and the original particles lose their individual identities.

DISCUSSION

By using a hemispheroidal particle as a prototypic IMP and computer simulation techniques, it is demonstrated that the dimensions of a particle's cap and shadow region do not have a simple relationship to actual particle dimensions. Shadowing conditions and the particle's geometry (plus other factors in actual practice) influence the particle's image. Our results suggest that if replica thickness is known, as is the case when a quartz film thickness analyzer is used in the preparation of freeze-fracture replicas, it may be possible to deduce approximate IMP dimensions and shapes, using synthetic particle images as a frame of reference. This is supported by our observation that the electron microscopic images of actual IMP in biological membranes resemble the computer-generated synthetic images and may be related to them through the use of the appropriate correction factors (2).

A problem in simulating electron optical images of real particles involves the formulation of correction factors for cap width. For the simulations described in this paper, two assumptions are made that tend to restrict the widening of the metal cap as the replica increases in thickness. The first assumption is that the atoms in the flux of replicating materials stick where they hit the specimen surface. The second assumption is that atoms at the lateral margins of the metal cap adsorb only replicating atoms from the flux if there is a direct hit. Atoms grazing the sides of the cap bounce off or pass through. The first assumption is open to debate (20), whereas the second assumption is an obvious simplification of what occurs at the growth front of a replica.

Studies by Hall (5) and Misra and Das Gupta (16) on specimens other than freeze-fractured membranes have shown that under certain conditions the width of the metal cap on particles can exceed the actual diameter of the particle (5,16). Empirical correction factors have been devised to account for cap widening (16) and these factors have been used to estimate the actual dimensions of IMP in freeze-fracture replicas (10). This may be unjustified since the observations upon which the correction factors are based were obtained under replication conditions that are very different from those encountered in freeze fracturing.

Both Hall (5) and Misra and Das Gupta (16) used platinum wire wrapped around a long tungsten filament as an extended unshielded source for their metal flux. An extended (elongated) source would tend to widen the cap by virtue of a dispersed incident flux of metal. Heating effects also may have contributed to cap widening in these experiments. Several factors may have contributed to specimen heating. Unshielded metal sources would result in the heating of the supporting surface by radiation. A hot surface increases the time that is required to dissipate the kinetic energy of the impacting atoms. The replicating atoms reevaporate under these conditions, with the replica surface then acting as a secondary source of replicating atoms (19). The situation may have been further aggrevated by the fact that the particles themselves were not cooled, unlike the situation in freeze-fracture replication where the IMP are at $-100°C$ or colder (17,18). The particles were round and rested upon a flat surface which adds to the inefficiency of heat dissipation since the surface area of the sphere in direct contact with the supporting surface would be minimal. Lastly, unshielded sources of replicating materials can produce ionic bombardment of the surface, causing separation of charge, replusion of similarily charged metallic ions, and atomic movement at the replica's surface.

Early freeze-fracture observations seemed to provide additional evidence that cap widening is an inevitable consequence of unidirectional shadowing (10,11,22). These were believed to provide a rationale for using the correction factor of Misra and Das Gupta (16) in the interpretation of freeze-fracture replicas. However, it emerged that a major contributor to the dimensional distortion of IMP in the early freeze-fracture replicas was fracture-free contamination *in vacuo* prior to and during specimen replication (9). Subsequent improvements in evaporator systems and in freeze-fracture techniques, and the introduction of the electron gun metal source have done much to eliminate the artifacts that were common in the early days of freeze-etching.

Current freeze-fracture methodology should help to minimize dimensional distortions of the replica, although it is unlikely that they can be entirely eliminated. The following factors help the situation: In freeze fracturing, the replicated surface is cold, reducing secondary source emission. The IMP are not isolated particles resting unattached on a support film but rather are mounds at an irregular fracture face. As a point in fact, they are not particles at all as conventionally defined in the thin-film literature but merely resemble particles because of their shadowing characteristics. Because they are a physical component of the underlying tissue, which, in turn, is fused to the cold specimen stage, heat dissipation takes place at a far greater rate than is possible with actual particles on a support film. On the other hand, since frozen tissue is itself a rather poor conductor of heat, some heating effects may persist in freeze-fracture replicas. The electron gun also helps to reduce artifacts. A magnet traps ions within the gun, preventing the new surface from becoming charged. A small aperture is used to reduce source size and radiant heating (7).

To summarize, the problem of cap widening in unidirectionally shadowed preparations probably has been reduced to some extent by improvements in technology. It remains to be proven if cap widening can now be handled with simple correction factors. In this regard, it may be noteworthy that comparative studies on the dimensions of oncornaviruses obtained by freeze fracturing, thin sectioning, and negative staining indicate that there is no widening of the images in freeze-fracture replicas (12).

A cohort effect is the influence of a particle (or other component of the surface being replicated) on the image of a neighboring particle. In unidirectionally shadowed replicas, a cohort effect is observed where two particles are in close proximity and their axes are approximately colinear with the shadowing source. The effect is created by the proximal particle's shadow overlapping the distal particle's metal cap. Such effects hamper the identification of individual particles and alter the apparent particle dimensions. Furthermore, they interfere with the measurement of the center-to-center spacing between particles since the apparent anatomical center of a particle is shifted. Correction factors for cohort effects may be approximated by reconstructions from simulated images in some unambiguous cases or they can be minimized through the use of rotary shadowing (6,13), in place of unidirectional shadowing. Rotary shadowing is an adequate solution to the cohort effect problem in some settings but is not universally applicable because particle heights are more difficult to measure in rotary shadowed replicas.[1] Particle height is a parameter that should be useful for subclassifying particles for numerical density and topographic analyses.

Another drawback with rotary shadowing relates to the area of membrane that is suitable for quantitative analysis of the IMP. The size of test fields that are suitable for quantitative analysis is severely limited in rotary shadowed replicas of curved membranes. This is because rotary shadowing is only uniform over 360 degrees where the membrane is coplanar with the specimen stage

---

[1] In theory, particle heights can be measured from rotary shadowed replicas since the gray level of circular metal cap, as recorded in electron micrographs, is related to cap height. Operationally, very few gray levels can be resolved in the electron dense metal cap region at the present time.

and perpendicular to the axis of rotation. For most biological specimens, larger areas of membrane are amenable to quantitation with unidirectional shadowing.

REFERENCES

1. Abermann, R., Salpeter, M. M., and Bachmann, L. (1972): High resolution shadowing. In: *Principles and Techniques of Electron Microscopy*, Vol. II, edited by M. A. Hayat, pp. 195–217. Van Nostrand Reinhold, New York.
2. Benefiel, D. J., and Weinstein, R. S. (1975): Image analysis of intramembrane particles in freeze-fractured biomembranes using computer simulation techniques. *Proc. 33rd Ann. Mtg. Electron Microsc. Soc. Am.* pp. 214–215. Claitor's Pub. Div., Baton Rouge.
3. Branton, D. (1969): Membrane structure. *Ann. Rev. Plant Physiol.*, 20:209–238.
4. Branton, D. (1971): Freeze-etching studies of membrane structure. *Philos. Trans. R. Soc. Lond.*. 261:133–138.
5. Hall, C. E. (1960): Measurement of globular protein molecules by electron microscopy. *J. Biophy. Biochem. Cytol.*, 7:613–618.
6. Heinmets, F. (1949): Modification of silica replica technique for study of biological membranes and application of rotary condensation in electron microscopy. *J. Appl. Phys.*, 20:384–389.
7. Henderson, W. J., and Griffiths, K. (1972): Shadow casting and replication. In: *Principles and Techniques of Electron Microscopy*, Vol. II, edited by M. A. Hayat, pp. 149–193. Van Nostrand Reinhold, New York.
8. Koehler, J. K., (1972): The freeze-etching technique. In: *Principles and Techniques of Electron Microscopy*, Vol. II, edited by M. A. Hayat, pp. 51–98. Van Nostrand Reinhold, New York.
9. Kreutziger, G. O. (1968): Specimen surface contamination and the loss of structural detail in freeze-fracture and freeze-etch preparations. *Pro. 26th Ann. Mtg. Electron Microsc. Soc. Am.*, pp. 138–139. Claitor's Publ. Div., Baton Rouge.
10. Lessin, L. S., Jensen, W. N., and Ponder, E. (1969): Molecular mechanism of hemolytic anemia in hemaglobin C disease. *J. Exp. Med.*, 130:443–466.
11. Lickfeld, K. G., Achterrath, M., and Hentrich,F. (1972): The interpretation of images of cross-fractured frozen-etched and shadowed membranes. *J. Ultrastruct. Res.*, 38:279–287.
12. Luftig, R. B., McMillan, P. N., Culbreth, K., and Bolognesi, D. P. (1974): A determination of the outer dimensions of oncornaviruses by several electron microscopic procedures. *Cancer Res.*, 34:1694–1706.
13. Margaritis, L. H., Elgsaeter, A., and Branton, D. (1977): Rotary replication for freeze-etching. *J. Cell Biol.*, 72:47–56.
14. McNutt, N. S., and Weinstein, R. S. (1970): The ultrastructure of the nexus: A correlative thin-section and freeze-cleave study. *J. Cell Biol.*, 47:666–688.
15. McNutt, N. S., and Weinstein, R. S. (1973): Membrane ultrastructure at mammalian intercellular junctions. *Prog. Biophys. Mol. Biol.*, 26:45–101.
16. Misra, D. N., and Das Gupta, N. N. (1965): Distortion in dimensions produced by shadowing for electron microscopy. *J. R. Microsc. Soc.*, 84:373–384.
17. Moor, H., Mühlethaler, K., Waldner, H., and Frey-Wyssling, A. (1961): A new freezing-ultramicrotome. *J. Biophys. Biochem. Cytol.*, 10:1–13.
18. Nickel, E., and Grieshaber, E. (1969): Elektronenmikroskopische Darstellung der Muskelkapillaren im Gefrierätzbild. *Z. Zellforsch.*, 95:445–461.
19. Preuss, L. E. (1965): Shadow casting and contrast. *Lab. Invest.*, 14:919–932.
20. Ruben, G.G., and Telford, N. J. (1975): The platinum–carbon shadow width increase as a function of known particle size on non-planar surfaces. *Proc. 33rd Ann. Mtg. Electron Microsc. Soc. Am.*, pp. 282–283. Claitor's Publishing Div., Baton Rouge.
21. Steck, T. L. (1974): The organization of proteins in the human red blood cell membrane. *J. Cell Biol.*, 62:1–19.
22. Weinstein, R. S., and McNutt, N. S. (1970): Electron microscopy of red cell membranes. *Semin. Hematol.*, 7:259–274.
23. Zeitler, E., and Bahr, G. F. (1965): Contrast and mass thickness. *Lab Invest.*, 14:946–954.

Freeze Fracture: Methods, Artifacts, and Interpretations, edited by J. E. Rash and C. S. Hudson. Raven Press, New York © 1979.

Interpretation of Freeze-Fracture and Freeze-Etch Images: Morphology and Realism

Pedro Pinto da Silva

Section of Membrane Biology, Laboratory of Pathophysiology, National Cancer Institute, National Institutes of Health, Bethesda, Maryland 20205

Linha severa de longínqua costa—
Quando a nau se aproxima ergue-se a encosta
Em árvores onde o Longe nada tinha;
Mais perto, abre-se a terra em sons e cores.
E, no desembarcar, há aves, flores,
Onde era só, de longe a abstracta linha.

Fernando Pessoa, Horizonte (17)[1]

Freeze fracture and freeze etching have provided us with a new view of biological membranes. When compared with other electron microscopic methods, these techniques are well grounded in physical chemical principles. It is no exaggeration to say that freeze fracture has produced the most solid evidence for current concepts of the structure of biological membranes. As freeze-fracture and freeze-etch morphologists, we are aware that its capacity to show nonaverage images of membranes is unique. This contrasts with most physical and biochemical techniques which, able to provide only averaged views, are impotent in face of the noncrystalinity generally displayed by the planar distribution of membrane components.

In freeze fracture (32), the solutions of technical problems lead to satisfactory replicas of cellular membranes (2,15). Initially, however, the fracture plane was thought to follow membrane surfaces. Later, Branton hypothesized that, instead, fractures split membranes along their hydrophobic interior (3). At stake in determining the location of the fracture plane were not only important but narrower issues of interpretation of a new technique, but also basic questions on the structure of biological membranes.

To prove membrane splitting, it became necessary to observe a freeze-fractured

---

[1] Simple line of distant coast—
Galleon approaches: hillsides loom
And trees rise from the void of far.
Closer, the earth bursts in sounds, in colors
And, disembarking, there are birds, flowers
Where it was, from far, line abstract.

FIG. 1. Fractured and etched erythrocyte ghost shows ferritin only on etched outer surface. ×55,000.

and etched biological membrane after labelling of both outer and inner surfaces with a covalently attached, morphologically unequivocal marker, to show that the label could not be observed on either the A (P) or B (E) fracture faces (Fig. 1) (19,20).[1] These experiments were necessary, notwithstanding arguments in favor of membrane splitting which indicated that a natural cleavage plane could be provided by noncooperative van der Waals interactions formed at the hydrophobic juncture of acyl chains in a bilayer. Indeed, it could be argued that a natural cleavage plane could, instead, form during freezing by an advancing front of hydrogen bonds stopping close to the membrane surface where water molecules might be organized into an H-bond network and, thus, might not be available for further propagation of the freezing front.

About this time, Robert Waaland, who was studying gas vacuoles (36), thought of mixing erythrocytes with collapsed and uncollapsed gas vacuoles (Fig. 2) and observing the results by freeze fracture and freeze etching. To our astonishment (and no lesser aprehension!) he found that gas vacuoles, which had somehow adhered to the erythrocytes (or been pushed against them during freezing) were present on fracture face A (P) of the erythrocyte (Figs. 3 and 4). When

---

[1] Labeling of the outer surface alone (33) was insufficient to determine the location of the fracture plane. Equally insufficient was to find complementarity of A (P) and B (E) fracture faces (38). Proof of splitting applies only to the bilayer regions of the membrane; the exact location of the fracture plane at the particles was not, and is not, established.

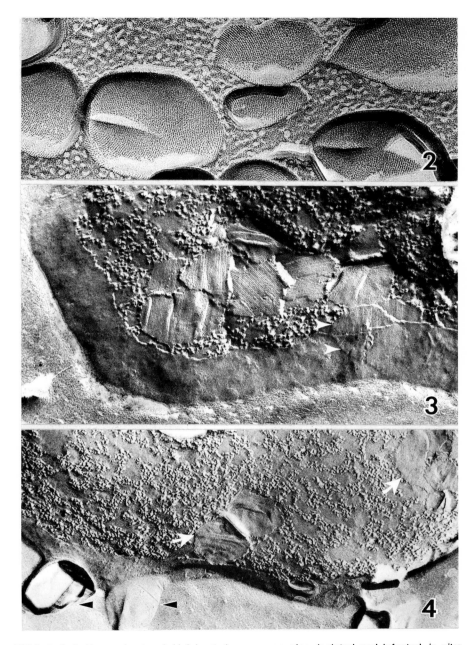

FIGS. 2–4. 2: Freeze-fractured *Halobacterium* gas vacuoles, isolated and infected *in vitro*. ×108,000. **3 and 4:** Erythrocyte ghost gas vacuole suspension, fractured and etched. **3:** Collapsed gas vacuoles are seen on the fracture face and also, less clearly, on the etched surface *(arrowheads)*. ×81,000. **4:** Uncollapsed gas vacuoles are also seen on the fracture face *(white arrows)* and attached to the outer surface *(arrowheads)*. ×63,000. (FIGS. 2–4 courtesy of Dr. J. R. Waaland.)

the preparations were etched, gas vacuoles could also be seen on the etched outer surface. To make matters worse, the same collapsed gas vacuole could be observed with one part on the fracture face, the other on the etched surface (Fig. 3, *arrows*). The obvious interpretation of these results was that, because a surface marker could be observed on a fracture face, the membranes were not split. In consequence, the plane of fracture had followed either the interface between the outer surface of the membrane and the surrounding medium; or, external to that, a cleavage plane was established during freezing between the advancing ice front and the organized layers of water at the outer surface. These were, indeed, days of uncertainty and puzzlement. Later, these results were clarified by a simple explanation: when the gas vacuoles collapsed against a membrane, their apposed walls formed a "bilayer." Because gas vacuoles are thought to be composed of a single protein (9,37) it would be unlikely that attractive forces of any kind might be established between their apposed inner surfaces. If anything, repulsive forces might exist between identical faces of the same protein. Thus, when the fracture plane, presumably following the membrane interior, reached a region containing gas vacuoles at the surface, it would choose the plane of least resistance, now provided by the apposed layers of the gas vacuole. One is led to imagine how such "convincing" results would have delighted the proponents of surface-fracture alternatives.

Proof of splitting of the bilayer region of the erythrocyte membrane and the qualitative similarity of all freeze-fractured biological membranes (smooth regions interrupted by particulate components) as well as indirect arguments on the chemical nature of the particles, led to the concept of the biological membrane as a bilayered continuum composed of lipid and adsorbed proteins locally interrupted by intercalated and/or sequestered protein-containing structures (20). This view of the membrane provided a synthesis of the antithetic concepts of membrane structure then prevailing, namely, that of Danielli and Davson (6) and Robertson (29) which viewed the membrane as an uninterrupted bilayer, and that of Sjostrand (30) and others which considered the membrane as a two-dimensional aggregate of globular lipid–protein subunits.

Proof of the proteinaceous nature of the particles and, equally important, demonstration that, at least in the erythrocyte membrane, they represented structures intercalated into the hydrophobic matrix of the membrane and exposed at the outer surface came in the following years through the use of membrane markers to locate antigen sites of the AB (H) system (21,22) and lectin and virus binding sites (34,35). The use of cationized ferritin to label anionic sites at the surface of right-side-out and inside-out erythrocyte membranes illustrated the transmembrane nature of the structure represented by the erythrocyte membrane particle (27). Implication of the components represented by the particles in transmembrane functions followed. Gap junction particles were found to represent structures of low electric resistance (8,14,16,23). Particles in sarcoplasmic reticulum membrane were shown to represent sites of ATPase activity (7,12). In human erythrocyte membranes, the particles were related to the sites that bound concanavalin A and, in consequence, to Band

III, a major integral protein implicated in the transmembrane passage of anions
(28).

Although the erythrocyte membrane was the most commonly used model
in early freeze-fracture and freeze-etch studies of membrane structure, it is clear
that it represented a relatively simplified model of biological membranes. This
was particularly important in studies of the mobility of membrane proteins
and the nature of the cell surface. In the erythrocyte, the main proteins expressed
at the outer surface are integral proteins (glycophorin and Band III components),
that are represented by the membrane particles (31). In consequence, movement
of surface antigens and other binding sites is the result of translational displacement of the membrane intercalated particles (18). In other cells, however, the
situation is more complex because, in addition, there is a wealth of peripheral
components that may move at the outer surface in a rather independent manner.
This was initially shown in lymphocytes (10) and in *Entamoeba histolytica*
(24,25), and the concepts of integral and peripheral fluidity of membranes advanced to account for the mobility of integral and peripheral membrane
components.[1]

Another and important difference between erythrocyte and other plasma membranes concerned the visual appearance of the "particle." In erythrocytes, all
particles are visually well-defined structures; but in most membranes, clear "particles" coexist with a multitude of less clear structural differentiations, rugosities
or "subparticles" (13,24). These structures are not always seen because they
are obliterated by slight amounts of contamination and generally best observed
in those regions of the fracture faces that received shadow at a lower angle
(26). These differences in the visual appearance of the particles may reflect
not only the chemical diversity of integral membrane components but also diversity in mode and extent of apolar stabilization with neighboring molecules.
Thus, it may be premature to consider particles in general (i.e., subparticles
included) to be intercalated, protein-containing, structures. Indeed, any generali-
zation must, at present, be based on a relatively small number of membranes
and direct evidence is lacking for the composition and location of the membrane
particles of most plasma membranes as well as intracellular membranes.

Questions of interpretation have been a constant companion of the analysis
of freeze-fracture and freeze-etch images; but definitive solutions of some (e.g.,
proof of splitting of the bilayer regions of the membrane) tend to mask the
existence of many other still unresolved questions that challenge the limits of
our ability to perceive the visual image. Although firm answers to interpretative
problems form a platform for routine analysis of freeze-fracture and freeze-
etch images, a continued need remains for enlarging the domains accessible to
interpretation. This necessarily involves the acceptance of subjective interpreta-

---

[1] It is clear that "integral" and "peripheral" represent extreme situations and that components
with characteristics of both exist. For instance, the basic protein of myelin, exists at the cell surface
but segments of the molecule probably interact hydrophobically with other components of the
membrane (4).

tion, which is inherent in any perceptual process and immanent in the epistemological basis of morphology.

MORPHOLOGY AND REALISM

O sonho é ver as formas invisíveis
Da distância imprecisa, e, com sensíveis
Movimentos da esp'rança e da vontade,
Buscar na linha fria do horizonte
A árvore, a praia, a ave, a fonte
Os beijos merecidos da Verdade
        Fernando Pessoa, Horizonte[1]

The analysis of freeze-fracture and freeze-etch experiments is based on the observation of electron micrographs. With the solution of basic interpretive problems, it may seem that visual interpretation would play a lesser role than in other electron microscopic techniques where the images obtained appear less clear (e.g., negative staining of membranes) and require difficult analysis. But one of the advantages of freeze fracture—its beautiful and convincing images—could be converted into its nemesis if we mistake these powerful images of platinum casts of frozen and split membranes for those much more subtle and delicate structures which are the membranes themselves. In fact, what we see is a most subjective image, albeit masked by a deceiving crispness and beauty. And even though it is clearly necessary to strive for a greater understanding and control of each of the preparative steps, visual interpretation will remain a necessity. In addition, and to form within ourselves a more complete concept, our image of the membrane cannot remain in the realm of the visual but must be integrated with the contributions from other techniques, and also with our own empathy towards the membrane and, indeed, the cell.

It is clear that, as with any other technique, freeze fracture is subject to many artifacts. However, we must look at the state of our knowledge as an ever-changing and delicate interface between past and future understanding and, thus, continuously renewed. It is through the course of this process that we come closer to ever-clearer shadows of reality. This is the Aristotelean path of investigation: "To advance from what is clearer to us, though intrinsically more obscure, towards what is intrinsically clearer and more intelligible" (1). To go from reception into perception.

In the morphological sciences, the striving to obtain reflections closer to the living reality is both natural and legitimate. But reality is not and will not be objectively available. It then comes as a paradox to find among morphologists

---

[1] To dream: to perceive form invisible
From imprecise distance, and with sensitive
Movement of hope and of will
To search the cold line of the horizon
For tree, beach, flower, bird and spring
Deserved embrace with Truth.

such an obsession with reality when the concept of approximation is accepted and basic to the methods of the exact sciences.

Because all forms of creation seem to converge, art, the basic approach to science, and even humor find themselves on common ground (11). Thus, it is not surprising that the theme of this symposium, that is, the search for truth in a freeze-fracture image, finds parallels in the discussions about realism that pervaded literature and art at the beginning of this century. In art, the question was whether the visual copying of reality was acceptable, or whether better approximations might not be obtained by interpretive, subjective, and often symbolic means. Well over 100 years ago, Delacroix warned against those who sought visual replicas of reality, mistaking them for the truth (5).

Today, in art the issue has been settled, at least for the moment, in favor of those who hold a subjective view, and realism is considered to be only one of the subjective approaches to reality. In contrast, in science the issue seems to be resolved in the opposite direction and the lovers of "hard science," absorbed by those "irreducible, stubborn facts" (39) seem to carry the day. It is, of course, clear that practical requirements of science place limits on the subjective interpretation of reality. However, in biology the trend toward objectivism can also be accounted for by the types of laboratory techniques that have evolved and, in no small measure, by the golden era of the soluble-phase biochemists who, while discovering metabolic pathways, become used to disregarding the form in favor of the contents, viewing cells as nuclei, mitochondria, and, later, lysosomes, and "microsomes" suspended in a cytosol. Naturally affected by the progress of other "harder" sciences, we morphologists forgot a tradition of centuries and attempted to adopt the ways and, when possible, the mannerisms of the hard scientists. In doing so, we have evolved to a state in which, reduced to mere photographers, we may tend to forget the very basis of the morphological method, that is, the interpretation of form and the inference of function. Nevertheless, if it will always be necessary to obtain new images and closer reflections of a biological reality, we need not exclude *a priori* those, which, deformed and subjective as they may be, will also lead us to richer interpretations.

ACKNOWLEDGMENTS

I wish to thank Dr. J. R. Waaland for permission to describe his work on gas vacuoles, and Dr. Pietro Gullino, Dr. John Weinstein, Mrs. Nancy Dwyer, and Mr.Clifford Parkison for their helpful suggestions and support.

REFERENCES

1. Aristotle (1908–1952): Physics. Revised text with introduction and commentary by W. D. Ross. Oxford University Press, New York.
2. Branton, D. and Moor, H. (1964): Fine structures in freeze-etched *Allium cepa* L. root tips. *J. Ultrastruct. Res.,* 11:401–411.

3. Branton, D. (1966): Fracture faces of frozen membranes. *Proc. Natl. Acad. Sci. USA,* 55:1048–1055.
4. Braun, P. E., and Brostoff, S. W. (1977): Proteins of myelin. In: *Myelin,* edited by P. Morell, pp. 201–231. Plenum Press, New York.
5. Chierico, O. (1977): *Matilde Grant.* Editorial La Barca Grafica,Buenos Aires.
6. Danielli, J. F., and Davson, H. (1935): A contribution to the theory of permeability of thin films. *J. Cell Comp. Physiol.,* 5:495–508.
7. Deamer,D. W. (1973): Isolation and characterization of a lysolecithin–adenine triphosphatase complex from lobster muscle microsomes. *J. Biol. Chem.,* 248:5477–5485.
8. Gilula, N. B., Reeves, R. O. and Steinbach, A. (1972): Metabolic coupling, ionic coupling and cell contacts. *Nature,* 235:262–265.
9. Jones, D. D., and Jost, M. (1970): Isolation and chemical characterization of gas vacuole membranes from *Microcystis aeruginosa. Arch. Mikrobiol.,* 70:43–64.
10. Karnovsky, M. J., and Unanue, E. R. (1973): Mapping and migration of lymphocyte surface macromolecules. *Fed. Proc.,* 32:55–59.
11. Koestler, A. (1964): *Act of creation.* Macmillan, New York.
12. MacLennan, D. H., Seeman, P., Iles, G. H., and Yip, C. (1971): Membrane formation by the adenosine triphosphatase of sarcoplasmic reticulum membranes. *J. Biol. Chem.,* 246:2702–2710.
13. Martinez-Palomo, A., Pinto da Silva, P., and Chavez, B. (1976): Membrane structures of *Entamoeba histolytica:* Fine structure of freeze-fracture membranes. *J. Ultrastruct. Res.,* 54:148–158.
14. McNutt, N. S., and Weinstein, R. S. (1970): The ultrastructure of the nexus. A correlated thin section and freeze-cleave study. *J. Cell Biol.,* 47:666–688.
15. Moor, H., and Mühlethaler, K. (1963): Fine structure in frozen-etched yeast cells. *J. Cell Biol.,* 17:609–628.
16. Payton, B. W., Bennett, M. V. L., and Pappas, G. D. (1969): Permeability and structure of junctional membranes at an electronic synapse. *Science,* 166:1641–1643.
17. Pessoa, F. (1959): *Mensagem,* p. 108. Edições Ática. Lisboa.
18. Pinto da Silva, P. (1972): Translational mobility of the membrane intercalated particles of human erythrocyte ghosts. pH-dependent reversible aggregation. *J. Cell Biol.,* 53:777–787.
19. Pinto da Silva, P., and Branton, D. (1969): Freeze-etch studies of natural and artificial membranes using specific markers. Int. Botanical Congress Abstracts, Seattle p. 171. Aug. 23–Sept. XI 2.
20. Pinto da Silva, P., and Branton, D. (1970): Membrane splitting in freeze-etching. Covalently bound ferritin as a membrane marker. *J. Cell. Biol.,* 45:598–605.
21. Pinto da Silva, P., Douglas, S. D., and Branton, D. (1970): Location of A antigens on the human erythrocyte membrane. *J. Cell Biol.,* 47 (2, part 2):159a.
22. Pinto da Silva, P., Douglas, S. D., and Branton, D. (1971): Localization of A antigen sites on human erythrocyte ghosts. *Nature,* 232:194–196.
23. Pinto da Silva, P., and Gilula, N. B., (1972): Gap junctions in normal and transformed fibroblasts in culture. *Exp. Cell. Res.,* 71:393–401.
24. Pinto da Silva, P., and Martinez-Palomo, A. (1974): Induced redistribution of membrane particles, anionic sites and con A receptors in *Entamoeba histolytica. Nature,* 249:170–171.
25. Pinto da Silva, P., Martinez-Palomo, A., and Gonzalez-Robles, A. (1975): Membrane structure and surface coat of *Entamoeba histolytica.* Topochemistry and dynamics of the cell surface: Cap formation and microexudate. *J. Cell Biol.,* 64:538–550.
26. Pinto da Silva, P., and Miller R., (1975): Membrane particles on fracture faces of frozen myelin. *Proc. Natl. Acad. Sci. USA,* 72:4046–4050.
27. Pinto da Silva, P., Moss, P., and Fudenberg, H. H., (1973): Anionic sites on the membrane intercalated particles of human erythrocyte ghost membranes. Freeze-etch localization. *Exp. Cell Res.,* 81:127–138.
28. Pinto da Silva, P., and Nicolson, G. (1974): Freeze-etch localization of concanavalin A receptors to the membrane intercalated particles of human erythrocyte ghost membranes. *Biochim. Biophys. Acta,* 363:311–319.
29. Robertson, J. D. (1961): The unit membrane. In: *Electron Microscopy in Anatomy,* edited by J. D. Boyd, F. R. Johnson, and J. D. Lever, pp. 74–79. Williams and Wilkins, Co., Baltimore, Md.
30. Sjostrand, F. S., and Barajas, L. (1968): Effect of modification in conformation of protein molecules on structure of mitochondrial membranes. *J. Ultrastruct. Res.,* 25:121–155.

31. Steck, T. (1974): The organization of proteins in the human red blood cell membrane. *J. Cell Biol.*, 62:1–19.

32. Steere, R. L. (1957): Electron Microscopy of structural detail in frozen biological specimens. *J. Biophys. Biochem. Cytol.*, 3:45–60.

33. Tillack, T. W., and Marchesi, V. T. (1970): Demonstration of the outer surface of freeze-etched red blood cell membranes. *J. Cell Biol.*, 45:649–653.

34. Tillack, T. W., Scott, R. E. and Marchesi, V. T. (1971): Further studies on the association between glycoprotein receptors and intramembranous particles of the red blood cell membrane. *Abstracts 11th Ann. Mt. Am. Soc. Cell Biol.*, p. 305. Jung Hotel, New Orleans, La. Nov. 17–20, 1971.

35. Tillack, T. W., Scott, R. E., and Marchesi, V. T. (1972): The structure of erythrocyte membranes studied by freeze-etching. II. Localization of receptors for phytohemagglutinin and influenza virus to the intramembranous particles. *J. Exp. Med.*, 135:1209–1227.

36. Waaland, J. R., and Branton, D. (1969): Gas vacuole development in a blue-green alga. *Science*, 163:1339–1341.

37. Walsby, A. E. (1972): Structure and function of gas vacuoles. *Bacteriol. Rev.* 36:1–32.

38. Wehrli, E., Mühlethaler, K., and Moor, H. (1970): Membrane structure as seen with a double replica method for freeze-fracturing. *Exp. Cell Res.*, 59:336–339.

39. Whitehead, A. N. (1967): *Science and the Modern World.* The Free Press, New York.

Subject Index